ENERGY FROM THE SUN
—33 Easy Solar Projects

Dedication

To "JoJo" with love and appreciation.

No. 1323
$10.95

ENERGY FROM THE SUN
—33 Easy Solar Projects

by Isaac R. Holstroemn

TAB BOOKS Inc.
BLUE RIDGE SUMMIT, PA. 17214

FIRST EDITION

FIRST PRINTING

Copyright © 1981 by TAB BOOKS Inc.

Printed in the United States of America

Reproduction or publication of the content in any manner, without express permission of the publisher, is prohibited. No liability is assumed with respect to the use of the information herein.

Library of Congress Cataloging in Publication Data

Holstroemn, Isaac R.
    Energy from the sun—33 easy solar projects.

    Includes index.
    1. Solar energy—Handbooks, manuals, etc. I. Title.
TJ810.H65   621.47   81-9164
ISBN 0-8306-0023-X   AACR2
ISBN 0-8306-1323-4 (pbk.)

Contents

1 The Makeup of Light — 7
Wavelength—Light is Energy—Photoelectric Effect—Lasers—Summary

2 Solar Cells And Related Devices — 15
The Photovoltaic Cell—Concentrating the Light—Combining Solar Cells—Photoconductive Cells—Transistors—Phototransistors—Other Light-Sensitive Devices—Summary

3 Electronic Project Building — 41
Building Tools—Test Instruments—Soldering Procedures—Proper Soldering Techniques—Forming Building Habits—Special Techniques For Solar Cells—Shock-Mounting of Solar Cells—Mounting of Solid-State Components—Integrated Circuit Building Techniques—Obtaining Components—Cross-Referencing—The Experimenter's Junk Box—Keeping Track of Electronic Components—Summary

4 Electronic Solar Projects — 78
Troubleshooting—Photovoltaic Light Meter—Photoresistor Light Meter—Combination Light Meter—Three-Volt Power Supply From the Sun—Six-Volt High-Current Supply—Versatile Light-Controlled Switch—Electronic Alarm Clock—Light-Controlled Electronic Organ—AM Radio Booster—Solar-Powered FM Radio—Solar-Powered AM Radio—12-Volt Battery Charger for Half the Price—Amateur Radio Transmitter for 40 Meters—Code Practice Oscillator—Phototransistor Commercial Killer—Audible Light Meter—Light-Powered/Light-Controlled Pulse Oscillator—Solar-Powered Watch for Less Than $2—150 Volts from a 1.35-Volt Solar Cell Supply—Solar-Powered Two-Meter Converter for the Amateur Radio Operator—High-Current SCR Switch—Light-Controlled Automobile Finder—Light Modulator—Motorized Pinwheel—Light-Regulated Power Supply—Remote Light-Controlled Automobile Finder—Light-Controlled Holding Switch—Light-Powered Field-Strength Meter—Light-Controlled Rotary Switch—Variable-Voltage Solar Power Supply—Solar-Powered Television Receiver—Summary

Appendix A Schematic Symbols — 175

Appendix B Resistor Color Code Chart — 178

Index — 179

Preface

Solar power has been talked about for quite some time; but little was done about harnessing this energy until recently, when the declining supply of fossil fuels brought home the importance of energy from the sun. The solar cell has been with us for many, many years but it is an inefficient converter of light energy into electrical current. Recently, large corporations have been experimenting with these devices, and significant increases in conversion efficiency have been obtained. Solar cells may one day be competitive with more standard methods of generating electricity and, if development continues, solar electricity may well be a major new source of energy for the world.

This is more than a simple project book. In addition to the 33 circuits presented (designed to be built by the home experimenter with average tools and instruments), a lot of practical information is included about light and light energy. It is essential to know the answer to the question, "What is light?" before you can move ahead in attempting to use this source of energy for power and control applications.

A wealth of information is also provided on home building techniques which will allow you to get the most out of your electronic experiments with a minimum expenditure and waste of time. The author hopes you will find this book educational, informational, and enjoyable. It is also hoped that the reader will continue his or her experimentation with solar power and move on to even more technical aspects of this field. Whether you are a seasoned home experimenter, an agressive newcomer, or somewhere in between, you will find the scientific discussions, circuits, and step-by-step procedural information to be on a level which can be appreciated and applied.

Chapter 1
The Makeup of Light

Light is something we depend upon every day, but most of us really have no idea just what light is. There is no simple explanation which will completely define the entire makeup of light but, for the purposes of this book, a brief discussion of the characteristics of light is in order.

In medieval times most people thought that light was emitted from the human eyes and traveled to the object which was seen. With the science of optics came the first technical evaluation of light, how it behaved, and what its limitations were.

For the purposes of our discussion, we can consider light to be like radio waves. Radio waves are simply forms of energy which are transmitted through space. Likewise, light is energy which has an impact on everything it comes in contact with.

WAVELENGTH

Figure 1-1 shows a typical voltage curve of a radio wave. This drawing shows one complete cycle. At first, the polarity of the wave travels from zero to a peak positive value and then drops through zero to a peak negative value before returning to zero again. Radio waves travel through space at the speed of light, which is approximately 186,000 miles per second. The *frequency* of the radio wave is determined by how many cycles are completed within a one second period of time or within the distance light travels in one second. A radio wave at a frequency of 1,000,000 Hz (Hertz, or cycles per second) will complete 1,000,000 cycles in one second.

Wavelength is defined as the physical distance the radio wave travels to complete one cycle. Obviously, this distance will vary depending upon the frequency of the radio wave. In the above example, one cycle will occur in one-millionth of the distance that light will travel in one second. The actual distance can be computed by figuring how many feet light will travel in one second and dividing that value by the frequency of the wave, which in this case is one million cycles per second.

Wavelength at radio frequencies is more often defined in terms of meters rather than feet. These distances are maybe long; for example, it may take nearly 1,000 feet to travel the distance that would be covered during a single cycle at the 1,000 Hz frequency. Lower frequencies require an even greater distance to encompass a single cycle. Human beings need special instrumentation to detect this radiated energy because none of our senses are capable of indicating its presence. A radio receiver is designed to detect the radio frequency energy and to convert a portion of it into audio waves which we can detect with our auditory sense. The ears are natural radiation detection devices.

Under the same principle, as the frequency of a wave is increased a shorter distance is required to encompass one complete cycle. As the frequency increases even further, the wavelength becomes so short that it can be detected by another natural receiver contained in the human body, the eyes. When we can detect radiation with our eyes, the wavelength is said to be in the visible range of frequencies. The part of the frequency spectrum of sunlight which our eyes can see is called *visible radiant energy*, better known as *light*.

When frequencies are low enough to be heard by the human ear, they are called audio frequencies, or sound. As the frequency increases, it passes through the ultrasonic and supersonic portions of the spectrum. From there, frequency increase travels through the radio frequency spectrum and then on up through infrared rays (which are heat waves just below the human visible range).

Next on the ever-ascending frequency chart is the visible light portion of the spectrum, which is extremely narrow. It encompasses the multitude of colors displayed by a prism. Above the visible light range are ultraviolet rays, X rays, gamma rays, and cosmic rays.

It is interesting to note that the sun, the main producer of light for this planet, produces an electromagnetic wave output which covers all of the frequencies just discussed. Only a very tiny portion of the sun's output can be detected by the human eye. Some of the other frequencies may be detected in other ways. Ultraviolet rays cause the skin to tan. We do not see the light which heats the earth, at a lower frequency than visible light rays. But these infra red rays are detected by the human senses as heat.

Color detection is a rather complicated process. Almost everything we detect as color is reflecting a portion of the sun's rays while other portions of this visible light are absorbed. For example, a red sweater reflects the color red from the visible light spectrum and absorbs the greens, blues, and violets that are also a portion of this same spectrum. The color red lies near the bottom of the visible light spectrum; green,

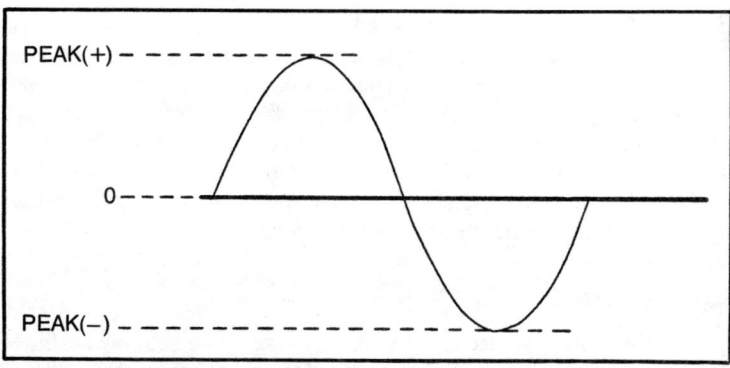

Fig. 1-1. Voltage curve of a radio wave.

blue, and violet are nearer the top. Any material which reflects the color red can be thought of as a low-frequency reflector and likened to a microwave radio reflector used for communications. The red sweater has been designed to communicate the color red to your eyes. The color which is transmitted is not that of the sweater but is a small portion of the sun's rays. This can be demonstrated by placing the same sweater in an enclosed room where no sunlight can leak in. By directing different colored incandescent lamps on the material, entirely different colors will be reflected back to the human eye. Remember, most of the objects you see do not contain their own coloration; rather, they receive electromagnetic input from the sun and reflect back a small portion to the human eyes.

Figure 1-2 shows the entire electromagnetic spectrum as we know it. Human beings make use of many portions of this spectrum. Some frequencies are used for communications, others for heating, and some (like X rays and gamma rays) for medical purposes. Cosmic rays lie at the upper end of the electromagnetic wave scale and come from outer space. Many scientists believe that cosmic waves are tiny atomic particles traveling at tremendous speeds, accelerated by the explosions of stars or entire galaxies.

When a material allows light to pass through it, it is called *transparent*. This means that a great portion of the visible light spectrum is allowed to pass unimpeded. A material that does not permit the passage of visible light is said to be *opaque*. *Translucent* materials scatter light.

It is necessary to understand some basic information about light before the operation of photoelectric devices such as solar cells can be fully explored. Light, as we now know, is the visible portion of all the electromagnetic energy released by the sun or by artificially manufactured sources (such as incandescent bulbs, fluorescent bulbs, etc.).

LIGHT IS ENERGY

All of us are familiar with the word *energy*; but again, many people do not know exactly what it means. *Energy* can be defined as the capacity for doing work. According to Pascal's Law, energy can neither be created nor destroyed. Therefore, much of the energy which went into the creation of visible light rays is carried with them; and this energy can be used if a way can be found to extract it. When a fire is built, the radiant energy transmits heat. When a radio wave is transmitted through space, some of the energy contained there can be sampled by a radio antenna and converted into current flow, another type of energy. Likewise, the energy contained in light can be extracted and used to perform work.

PHOTOELECTRIC EFFECT

Certain materials are able to extract energy from light rays. This is called the *photoelectric effect*. A portion of the energy which strikes these materials is transferred or transformed. The solar cell transfers the energy contained in light rays to energy in the form of electrical current. The solar cell is not generating power internally; it is simply a conversion device. It allows us to take advantage of available energy which heretofore was usually absorbed by the earth.

Photoelectric effects may take many different forms. Certain materials are changed chemically (transformed) when subjected to light energy. Photographic film and paper are good examples of this. As long as they are shielded from the visible portion of the sun's rays (or radiation from an artificial light source), they remain in one state. But when the visible light energy is allowed to strike their surfaces, chemical changes occur. The energy in light has been transformed to perform work.

Some of you may be surprised to learn that plants are one of the most obvious forms of natural storage batteries. For example, energy from the sun is stored in wood pulp and later may be released in the form of radiant energy when the wood ignited. Oceans store energy by constantly replenishing hydrogen through photochemical processes.

The earth depends upon the sun not just for its warming effect, but also for supplying energy at many different frequencies, which may be stored by different natural life forms. Photoelectric devices may take advantage of part or all of the visible light frequency spectrum. Some devices will respond more to the lower frequencies of red than the violet spectrum. Others will respond in a reverse manner. These devices will convert more energy from the portion of the visible light spectrum they best respond to than from other portions. Most will work equally well with artificial light as with natural sunlight.

The photoelectric effect and its uses can best be appreciated by considering the two fundamental properties of light. The first of these states that light is a form of energy which is conveyed through space at an extremely high velocity of about 186,000 miles per second (300,000 kilometers per second). This form of energy transmission is significant when compared to other forms of energy, such as chemical energy stored in oil, coal, wood, and other fissionable materials. Light energy is always

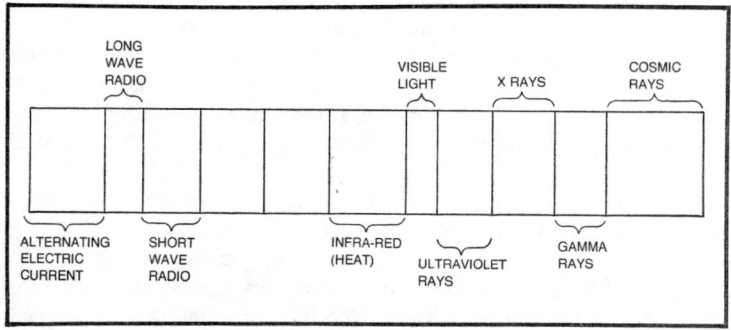

Fig. 1-2. The electromagnetic spectrum.

moving; it is never stationary. The movement of light is usually unaffected by the materials it strikes or passes through, although portions may be reflected away or bent from the initial path. All movement ceases when all of the energy has been absorbed. When this occurs, light is no longer present because the energy that was absorbed was the light itself. Remember, energy can neither be created nor destroyed; the light energy was not used up, it was simply transferred to another form such as radiant heat or current flow.

The second fundamental property of light is its ability to convey information. Scientists use this ability in astronomy to determine the chemical makeup of stellar bodies, to determine the heating effect of distant stars, etc. The information that light carries provides a wealth of data about the source of the light and also about the materials which tend to absorb this energy.

Photoelectric devices take advantage of both of these fundamental properties of light and put them to use in many different ways. Looking at the first fundamental property, the solar cell extracts energy from light, producing electricity which is used to power electronic circuits. Other types of photoelectric devices take advantage of the information properties of light rays by changing their internal resistance upon being subjected to light energy in pulsed form. The human voice may be sent and detected by means of light rays. The audio information of the voice is superimposed upon a generated light ray; the light intensity fluctuates with the voice frequencies. This fluctuation contains the information needed for detection at the receiving end. The beam of light is transmitted through space until it strikes a photoelectric substance which reacts directly to the information contained in the light ray. This may be a solar cell or some other type of photoelectric device. Information is extracted from the light ray and converted back into sound. This may sound like a complicated process, but it can be closely likened to the child's tin-can telephone made with two cans and a string. When the string is pulled taut between the two cans, a person speaking into the can at one end is heard in the can at the other end. In this case, the string could be likened to the light ray. When audio information is directed into the can, it creates a vibration or varying wave which travels along the string and sets up identical vibrations in the other can, which converts this physical vibration back into audio information. Light-wave transmitters and receivers are not all that complicated. Light-wave walkie-talkies have even been offered as toys by mail order catalogues.

Most of the projects in this book depend on sunlight or incandescent light for power or control. Smaller projects which receive power through solar cell conversion can receive energy from the sun or from artificial lights; but the larger projects which utilize many solar cells will most likely require the intense light which can be provided only by the sun. Some projects will be powered by batteries or other conventional sources and will depend upon light deviation to control volume, frequency, etc. These latter circuits will require only small light sources which may be in the form of flashlights, panel lights, or even light-emitting diodes.

Since light can be directed with mirrors and focused with proper lenses, the reader may wish to experiment with these implements in order to see what effects these devices could have on the operation of the electronic circuits to be built later. By knowing just what light is, it is easier to set up experiments which have some direction and meaning. Certain types of filters are available (through hobby and photographic outlets) which will allow experimental control by using light rays which are invisible in certain settings and under special conditions. It is interesting to observe how some of the photoelectric devices respond to different frequencies of light.

In experimenting with solar cells, it will be learned in later chapters that concentration of the light rays is very important. As a matter of fact, scattering of light is a major problem in maintaining strict control of photoelectric circuits. In addition to building the circuits contained in this text, there are possibilities for in-depth studies and experiments with lenses, filters, and direct-focusing devices to control the light beam which is ultimately used to power or to initiate the actions in these projects. Some experimenters have even resorted to low-power lasers for communciations and control purposes. Laser devices are available on today's electronic market for fairly reasonable prices, and they offer many possibilities for photoelectric experimentation.

LASERS

An electromagnetic wave is composed of excited atoms. For example, in an incandescent light bulb, electric current flowing through a conductor filament causes heating. The atoms within this filament begin to move and quickly reach an excited state. In a fraction of a second, each individual atom unloads its energy as a minute pulse of light. Each atom acts independently, ignoring the light path of each of the many other billions of atoms.

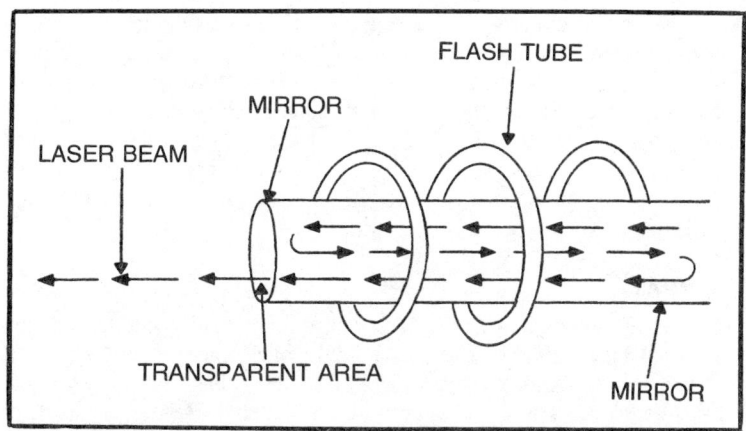

Fig. 1-3. Basic diagram of a laser.

Since each atom acts independently, the overall light radiation from an incandescent bulb is not orderly. A normal incandescent light bulb filament contains many different kinds of atoms and each type reaching an excited state at a different rate. Each type will also radiate at a different light frequency.

A *laser* produces *coherent* light. This means that the light produced is orderly, with all of the excited atoms making up the laser beam emitting their flashes in unison. The laser often consists of a polished ruby rod having a solid mirror at one end, transparent mirror at the other end, and a high-voltage flash tube wound around the rod. The flash tube acts as the power source for the laser beam, the energy being emitted as pulsed, visible light (see Fig. 1-3).

The flash tube emits brilliant pulses of light that stimulate the electrons in the ruby atoms. They, in turn, emit light which is then reflected between the mirrors in the rod. This further stimulates the ruby electrons to generate more light. The result is that the light which emerges in parallel waves through a small transparency in one mirror is many times more intense than that of the flash tube which initiated the action.

Through solid-state technology, laser diodes have been developed which are semiconductor diodes usually made from specially treated gallium. When a voltage is applied to the terminals of a laser diode, a coherent light beam is emitted.

The energy contained in a coherent light beam is highly concentrated and may often be applied directly to work situations. Lasers have been developed which will cut through steel or weld heavy metals. The concentration of energy contained in the light pulses through a coherent transmission system found in the laser makes this direct usage of light energy not only possible but practical as well. Many industries now use lasers for delicate cutting and welding work.

Electronic circuits have been designed which divide the transmitted laser beam into pulses. Pulses must follow in close procession, but each must also remain distinct. One system has been developed, using laser crystals, which allows a series of short pulses to be transmitted (each of which begins and ends within 30 billionths of a second). These pulses are transmitted at a rate which corresponds to the audio input at the modulator portion of the circuit and can later be returned to an audio state at the receiving end.

SUMMARY

Light is energy; and we are constantly bombarded by energy from the sun and from other stars in the universe. Light rays are constantly moving and when they exist no longer, all of their energy has been absorbed and transferred into a new energy system. Photoelectric devices are made of materials which have the capability of transforming light energy into electrical energy, or which go through chemical or compositional changes when absorbing visible light.

Most photoelectric devices respond to visible light, just as is the case with the human eye. Some devices act as energy sources when being fed by the sun's rays or light from an artificial source, while others utilize conventional power sources and serve as electronic control devices whose actions are stimulated by the presence, absence, intensity, or frequency of light waves.

Think of light in much the same way you think of radio transmissions. Photoelectric devices are similar to radio receivers which are directly affected by the transmissions they receive. By thinking of light in this manner, it will be much easier to understand the workings of photoelectric devices and to experiment further with the control of electronic functions and circuits through control of intensity, path, and travel of light rays.

Chapter 2
Solar Cells And Related Devices

When someone refers to a solar cell, this usually means a device which converts sunlight into voltage; but there are many other devices which also fit into this family more accurately known as the *photoelectric cell*. All of them have one thing in common—for a specific light input, a specific reaction occurs. This reaction will parallel the light input. Photocells, then, may also be called *transducers*. A transducer is a device which transfers power from one system to another. The common audio speaker is an excellent example of transduction, because it takes power from one system (electrical current flow) and transfers it to another (audio output). Photocells have light rays from the sun or other sources as their inputs. Their outputs, however, may differ considerably, depending on the device.

What most people refer to as a solar cell is more accurately known as a photovoltaic cell. It is made up of photovoltaic material which generates a voltage when exposed to light waves. The principle substances exhibiting this effect are selenium, germanium, and silicon. Silicon is almost always used in modern solar cell construction. Figure 2-1 shows a representative circuit which describes what a photovoltaic or solar cell does. The resistor represents an electrical load. When light strikes the surface of the cell, current begins to flow in direct proportion to the light intensity and frequency. The flow is from the negative pole of the cell to the positive. The solar cell, then, is a polarized device just like a dry cell battery. For this reason, the solar cell may be used as a direct battery substitute, providing load demands do not exceed the ratings of the solar cell. The photovoltaic cell transfers the power from light rays into electrical power. It is a transducer, because it transfers power from one system to another.

Another device which we will be working with throughout this book is known as a *photoconductive* cell. This is often called a photocell when sold

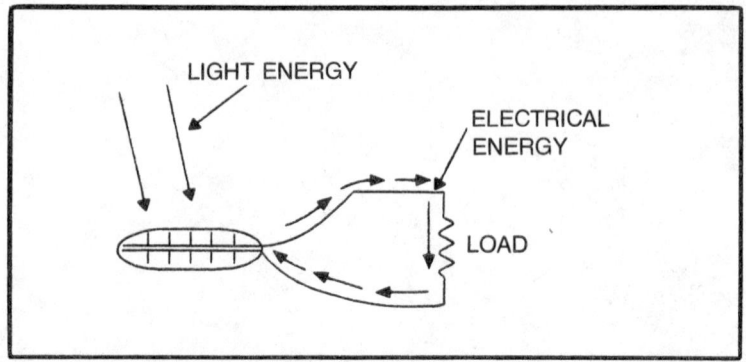

Fig. 2-1. Operation of a photovoltaic cell.

through hobby stores and should not be confused with the photovoltaic, cell which is more often called a solar cell. The operation of the two is completely different.

The photoconductive cell is made from cadmium sulfide in most instances and changes resistance in direct proportion to the intensity of the light which strikes its treated surface. The cadmium sulfide (CdS) cell, then, is a device whose resistance is variable according to the amount of available light. In practice, these cells exhibit decreased resistance when light is applied to their surfaces. The resistance begins to rise as the light intensity is lowered. We can say that the resistance is inversely proportional to the light intensity, while the conductivity is directly proportional to this same intensity. As the light intensity increases, so does the conductivity.

Figure 2-2 shows a representative circuit of a photoconductive cell's operation. Notice that a separate power supply is needed, as the CdS cell supplies no electrical power of its own, as was the case with the photovoltaic cell. In this circuit, the current flow is controlled by the internal resistance of the CdS cell, but the current source is a separate supply—in this case, a dry cell battery.

In the earlier circuit (Fig. 2-1), light controlled the current flow in the photovoltaic circuit. Light still controls current flow in the photoconductive circuit by determining the amount of resistance that flow will encounter. The only disadvantage with the latter circuit is the fact that, sooner or later, the battery must run down. The CdS cell will continue to change its conductivity as long as the light level changes; but after the battery is discharged, this resistance change will no longer have any effect on the dead circuit. Using the photovoltaic cell, the circuit is only active when light is striking this component's surface. When the light is gone, so is the control. The photovoltaic circuit will never run down, as long as there is some form of light present.

Just as the photovoltaic cell is often called by another name, so is the CdS cell, or photocell. A popular name for this component is photoresistor. Photoconductor would be as appropriate, but the former name seems to be the one which enjoys the most popularity.

Both of these devices operate as they do because of the *photoelectric effect*. This is defined as the phenomenon where temporary changes occur in the atoms of certain substances under the influence of light. Some of these materials undergo a change in their electrical resistance, as does the CdS cell; while others generate a voltage, as is the case with the photovoltaic cell.

While the photovoltaic cell is an electrical current device having a positive and negative pole, the photocell or photoresistor is non-polarized and may be treated like a carbon resistor for all intents and purposes. It is, however, a solid-state device and is not nearly as resistant to damage from current overloads as is the resistor.

THE PHOTOVOLTAIC CELL

Figure 2-3 shows a circular photovoltaic cell. It is made by combining two ultra-thin layers of silicon crystal which has been treated with certain impurities. To express the complicated construction process in simple terms, it can be said that one material is negative while the other is positive. When these two materials are sandwiched, one atop the other, a *p-n* junction is formed. This is where the photoelectric effect takes place. This phenomenon, discussed earlier, is the absorption of photons to create equal numbers of positive and negative charges.

Figure 2-4 shows a pictorial diagram of the construction of a photovoltaic cell. The top portion is the negative connection point, while the back surface is the positive pole. Solar cells are very fragile, as they are built on thin glass layers. They can be easily broken by mishandling.

From a practical aspect, we need to know more about what a solar cell does than how it is constructed. Generally speaking, the output current of

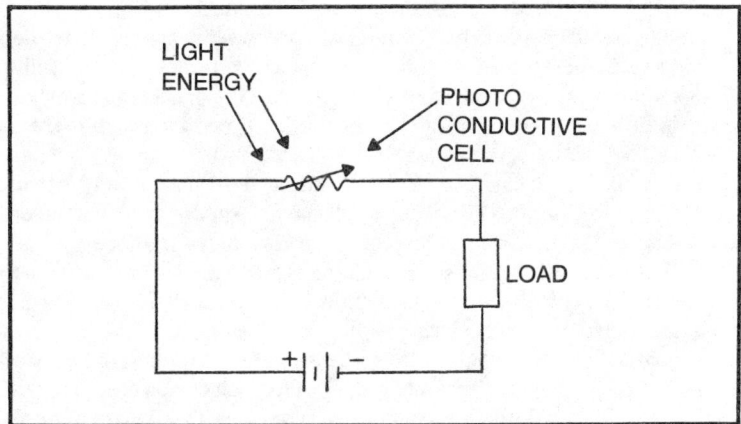

Fig. 2-2. Operation of a photoconductive cell in an electronic circuit.

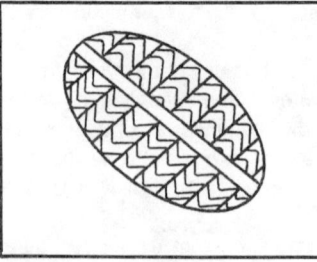

Fig. 2-3. A circular photovoltaic or solar cell.

a solar cell is directly proportional to the amount of light and the surface area of the cell. Assume that two cells, one with twice the surface area of the other, are constructed in the same manner. Given this set of circumstances, the larger cell would produce twice the current of the smaller one, assuming both are subjected to the same light levels. While output current will vary with the device size, voltage usually will not. Most solar cells produce a voltage in the area of 0.45 VDC. This may vary by a tenth of a volt due to differences in device construction from manufacturer to manufacturer. This voltage is the potential difference at the p-n junction. To get more voltage than this will require that the solar cells be combined with others in a series circuit. This will be discussed later.

While we might consider the power which is derived in the electrical circuitry formed by a solar cell as being free of charge, we must take into account the fact that the device itself costs something. While the prices of solar cells have dropped tremendously over the last five years, they are still quite expensive for the amount of power they ultimately deliver to the load. Solar cells are very inefficient. Of the total light energy which strikes the surface of a single cell, only about nine- to twelve percent of that power is delivered or transferred to the electrical output. The rest is reflected back into the atmosphere or lost through other channels.

Solar cell efficiency is improving, and devices which are claimed to be nearly twenty percent efficient are now considered to be economically feasible within the next decade; but, at present, the price tag is a major consideration. For example, a local hobby store carries a high-current solar cell which delivers one ampere of current at 0.45 volts DC under conditions of bright sunlight. This amounts to less than a half watt of electrical power per cell. Each solar cell carries a price tag of just under ten dollars. This breaks down to a cost of nearly $20 per watt of electrical power. The cell under discussion is not overpriced when compared to the rest of those available on today's market. All such devices are and will continue to be rather expensive.

Fortunately, for most of the projects outlined in this book, the high-current cells will not be needed. Smaller, lower-powered devices may be used instead which carry price tags of under five dollars in most instances. This is good, because most electronic solar cells as primary power supplies will require that they be used in a series connection to deliver at least 1.35 volts. When individual cells cost ten dollars or more, an innocent hobby can suddenly become a monetary burden.

Solar cells are available in a variety of shapes and sizes. The most common variety is round, as was shown previously; but others may be square, rectangular, or even in the shape of half moons. Each will produce maximum power when pointed directly toward the sun or other light source. This is where another problem may occur. Obviously, the more directly an object is pointed toward the sun, the more light rays it will receive. This is fine as far as solar cell operation is concerned. The unfortunate part involves heating effects. The more directly the cell is pointed at the sun, the hotter it gets. As the cell body temperature increases, its efficiency decreases. What you end up with is a seemingly impossible situation: the cell needs light to function, but the more it gets, the hotter it gets; and the hotter it gets, the more poorly it functions.

Fortunately, it is not quite as bad as it might seem. Cell efficiency is not adversely affected to the point where it can be detected until the cell body temperature climbs to over one hundred twenty five degrees. If exposed directly to the sun on a hot day, temperatures can easily reach this and higher levels; but most of the time a breeze will be blowing and will tend to keep the solar cell temperature at efficient levels. Some experimenters have devised ways to keep solar cells cool by circulating water around their lower surfaces through direct-coupled heat exchangers. This arrangement is very complex and would not be practical for

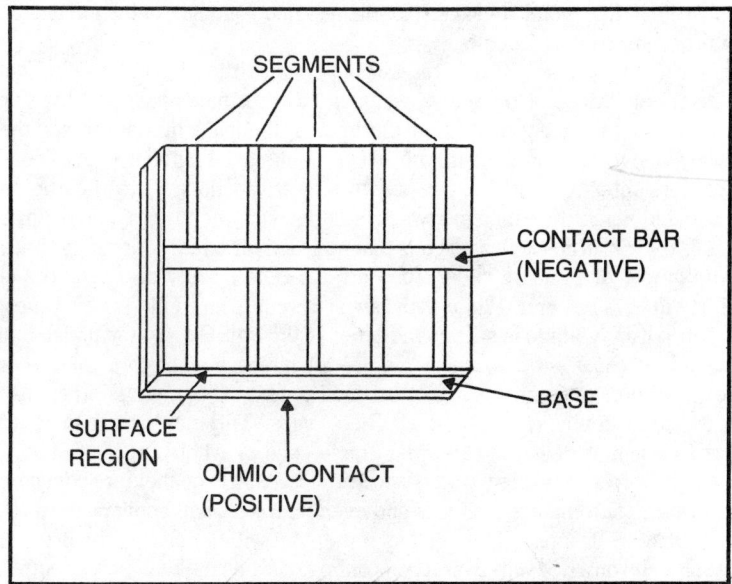

Fig. 2-4. Pictorial diagram of the construction of a photovoltaic cell.

any of the applications the solar cell is put to, in the following projects. If a bank of solar cells were used for the generation of electricity, then a cooling arrangement, controlled by a thermostat, might be in order. Generally, the temperature factor can be pretty much disregarded when solar cells are used to power experimental lobby projects like the many contained in this book.

CONCENTRATING THE LIGHT

It has already been stated that solar cells are inefficient for many reasons. One of these is the fact that much of the light which strikes the treated surface of the cell is reflected away. If you can concentrate the light on the cell surface in a more efficient manner, then the cell would have to produce more current. Figure 2-5 shows how this might be done using a simple hand mirror to reflect more of the sun's light rays back on the solar cell. Some experimenters have used fresnel lenses placed directly over the cell. This may produce up to thirty or forty times the natural light intensity.

It is difficult to arrive at a simple system for concentrating light on the surface of the solar cell. The rotation of the earth is a great hindrance to design simplicity because the sun will always seem to be moving out of alignment with the cell. It will be necessary to change the angle of the cell every few minutes to catch the maximum amount of light energy. Some solar power supplies actually track the sun. The cells and their concentrators are mounted on pivotal platforms driven by electric motors. A sensing device also made from photocells automatically *tracks* the sun and positions the pivotal stands so that the maximum amount of light available is concentrated on the cells.

The projects in this book will not require the elaborate setup just described. Most of them are designed to work under minimal lighting conditions; but for your future experiments, the many limitations and the ways of overcoming them are important points to be familiar with. Those readers interested in solar concentrators would do well to pick up an Edmund Scientific catalogue which features many different types of solar cells and solar concentrator cells that may deliver twenty- or thirty times the power of standard cells. Concentrator cells usually cost upwards of forty dollars per unit. While their power output is multiplied many times, their output voltage is still on the order of 0.5 volt DC. You will also find many accessory items which may be of interest when you take your experiments even further. Another good source of solar cells, other than your local hobby store, is Allied Electronics. Through their mail order catalog and stores, you can order single cells or whole panels, some of which deliver over thirty watts of power. These larger panels are adequate to charge automobile batteries and even operate small appliances which will take a DC input. If you go this route, you will notice that the price of solar cells on a per watt basis drops dramatically as the total power output of an array or bank of solar cells increases. This is because a large number

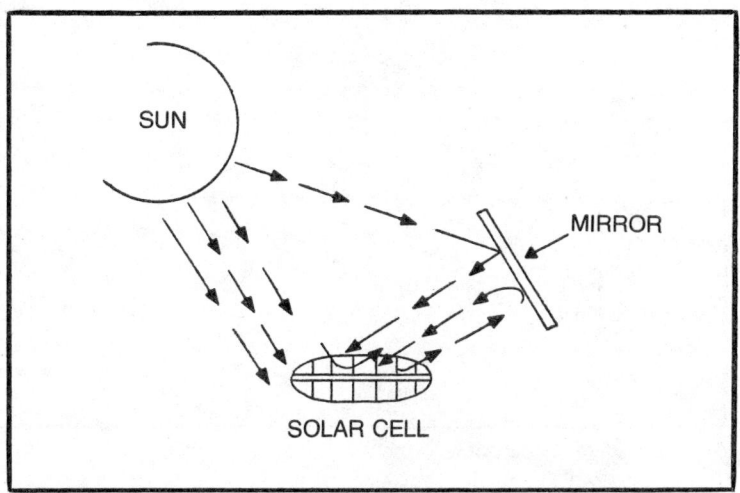

Fig. 2-5. Method of concentrating Sun's light on a photovoltaic cell.

of cells are combined to deliver this output, and you are getting a quantity discount of sorts.

COMBINING SOLAR CELLS

As was previously mentioned, the output from a solar cell is typically 0.45 volts, regardless of the nominal device current level. This is hardly enough to power even the tiniest transistor radio, so a means must be found to increase the voltage delivered to the electronic device to be powered. It was also stated earlier that the solar cell closely resembles the operation of a battery in that it has two poles, one positive and the other negative. As a matter of fact, a solar cell is also called a solar battery, although it does not store energy.

To increase the voltage when dealing with batteries, each component is wired in series with the positive pole of one attached to the negative pole of another, and so on. If we wire three batteries in a *series* circuit, then the total output voltage will be three times the value of any one, assuming that all three are identical. What about the current value in this series connection? It remains the same. If one battery will deliver 1.5 volts at 1 ampere, then three batteries in a series connection will deliver 4.5 volts at 1 ampere. Only the voltages add in a series circuit. The current rating remains the same.

Let's apply this to solar cells now. A single solar cell exhibits an output voltage of 0.45 volts DC. Figure 2-6 shows how three of them might be wired in series. Notice that the positive pole of one cell is connected to the negative pole of another which, in turn, has its positive pole connected to the negative pole of the third. The three components are wired in series just as if they were batteries.

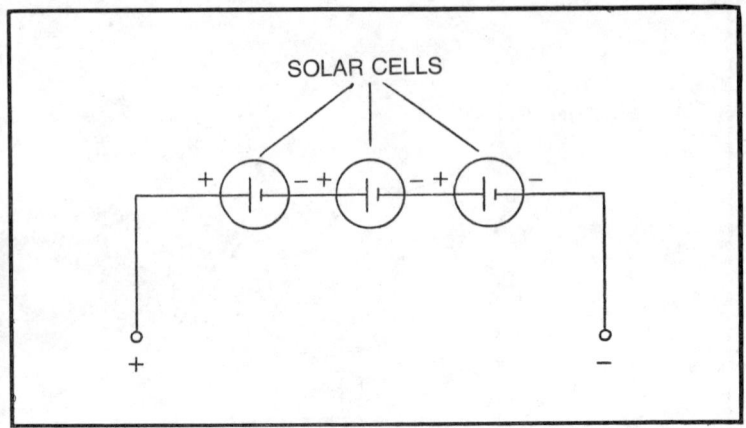

Fig. 2-6. Series connection of solar cells.

For the sake of discussion, we will say that each solar cell is capable of delivering 0.1 ampere to a normal load. Now that three have been connected in series, what will the voltage and current ratings be for this complex circuit? The answer is simple. In series circuits, the voltages add while the current remains the same. So, the output voltage will be equal to three times the value of a single cell, or 3 times 0.45. The output voltage will be 1.35 volts DC. The current rating will be the same as for a single unit, or 0.1 ampere.

By combining three cells in series, we have tripled the output voltage, kept the current capability the same, and tripled the power output over that obtainable with a single solar cell. Power is equal to the output voltage times the output current ($P = IE$). A single cell has a power output of 0.45 volt times 0.1 ampere, or 0.045 watts, or about one twenty-fifth of a watt. When three cells are combined, the power output rating is still voltage times current, but the voltage output has been tripled; so the formula reads Power = 1.35×0.1, or 0.135 watts, which is equivalent to about one-eighth of a watt.

Series connection is the rule rather than the exception with most solar cells used for power supply purposes. Often, you will find ten or more arranged in a series connection. Even a 6-volt supply would require at least thirteen cells. A 12-volt battery replacement would require that at least twenty-six solar cells be wired in series, and this is the case regardless of the power requirements of the load.

Again, for the sake of this discussion, let's assume that one solar cell delivers 0.45 volts at 0.1 ampere in bright sunlight. Assuming that the voltage output level presents no problem, how can we go about increasing the current rating? From the earlier example, we already know that three cells are capable of supplying about an eighth of a watt of power. Could we wire them in such a way as to have the output deliver the same 0.45 volts but at 0.3 ampere, three times the current of a single cell? Certainly. When solar cells are connected in series, the voltage of each unit adds to the total circuit voltage, but the current remains the same. However, when you wire the cells in *parallel*, the current adds while the voltage remains the same.

Figure 2-7 shows a circuit composed of the three solar cells. All three are connected in parallel with each other. Notice that the positive terminals of all three are tied together. The same is true of the negative terminals. Wired in this manner, the output voltage will still be 0.45 VDC, but the available current will be triple that of a single cell, or 0.3 ampere.

Now, what about the power output from this parallel circuit? Using (P = IE) for power again, we find that 0.45 times 0.3 still equals about one-eighth of a watt. This is the same power which was derived from the three cells in a series connection.

A lesson should have been learned by the examples given here. Each cell or battery is capable of delivering a specific amount of power. There is no way to increase this power. Additional cells may be added. With each cell, the power increases. Three cells can deliver three times the power of one. Thirty cells can deliver 30 times the power of one cell, and so forth. The power is more or less fixed by the number of components in the circuit, but the *manner* in which the power is *delivered* to the load may be changed and is dependent upon the method by which the many cells are wired in respect to each other and to the load.

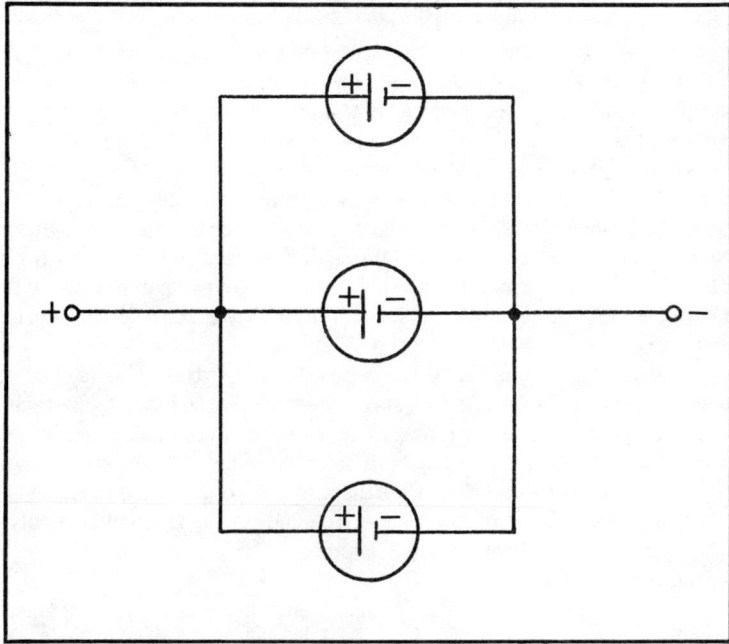

Fig. 2-7. Parallel connection of solar cells.

This may be a bit confusing to some readers. To put it more simply, the power will always add with each cell in the circuit. If one cell delivers one watt, then 20 cells will deliver 20 watts, regardless of whether the cells are connected in series or in parallel. In series connections, the voltage output of each unit is added, so 20 cells in series will deliver 20 times the voltage output of a single cell. Current remains the same and never adds in a series circuit. In a parallel circuit, 20 cells will deliver 20 twenty times the current of one cell, but the voltage never adds.

We *cannot* change the *total power* availability of a fixed number of cells, regardless of the manner in which we connect them. We *can* determine if the power available *can* be correctly used by the circuit under power by the manner in which the cells are connected. An electronic circuit may require that power be delivered at 1.35 volts in order for it to be used. We can see that the power is delivered in a usable manner by delivering it at 1.35 volts through a series connection. Another circuit may require its power be delivered at 0.45 volts and at 0.3 amperes of current. This delivery can be most efficiently made by a parallel connection. Both circuits have the same power capability. The trick is to deliver it in a way which can be used. The voltage and current ratings of the solar cell supply will determine this last factor, and the manner in which the cells are connected will determine the voltage and current ratings.

To take this discussion one step further, let's assume that our fictional electronic circuit needs power from the solar supply delivered at a value of 0.3 amperes at 1.35 VDC. How would this be accomplished? First, work the (P=IE) formula. Inserting the given values, we find that 0.3 times 1.35 works out to about three times the power rating of three of our sample solar cells. We already know that we cannot increase power without increasing the number of cells, so we must triple the number of cells to nine. This would give us approximately 0.4 watts of power, which is exactly what we need. But how do we combine the cells to equal 1.35 volts at 0.3 amperes? If we put all nine in series, then the resultant output voltage would be 9 times 0.45, or 4.05 VDC. The circuit requires only 1.35 VDC. On the other hand, if we combine all nine cells in parallel, we will have an output voltage still of only 0.45 VDC with a current of 9 times 0.1, or 0.9 amperes.

The answer to this riddle is a series-parallel circuit. This is shown schematically in Fig. 2-8. Three sets of three cells wired in series are then wired in parallel configurations. The result is three sections, each delivering 1.35 VDC at 0.1 amperes in the series configurations. These might be called series blocks. When these blocks are connected with each other in parallel, the current adds and we finally arrive at a power supply which delivers a total output of 0.3 amperes at 1.35 volts. This is exactly what the circuit required.

If the circuit needed higher voltage, then more cells would have to be added in series to increase the output. In doing this, the current capacity of the power supply would remain the same, because cells are only added in

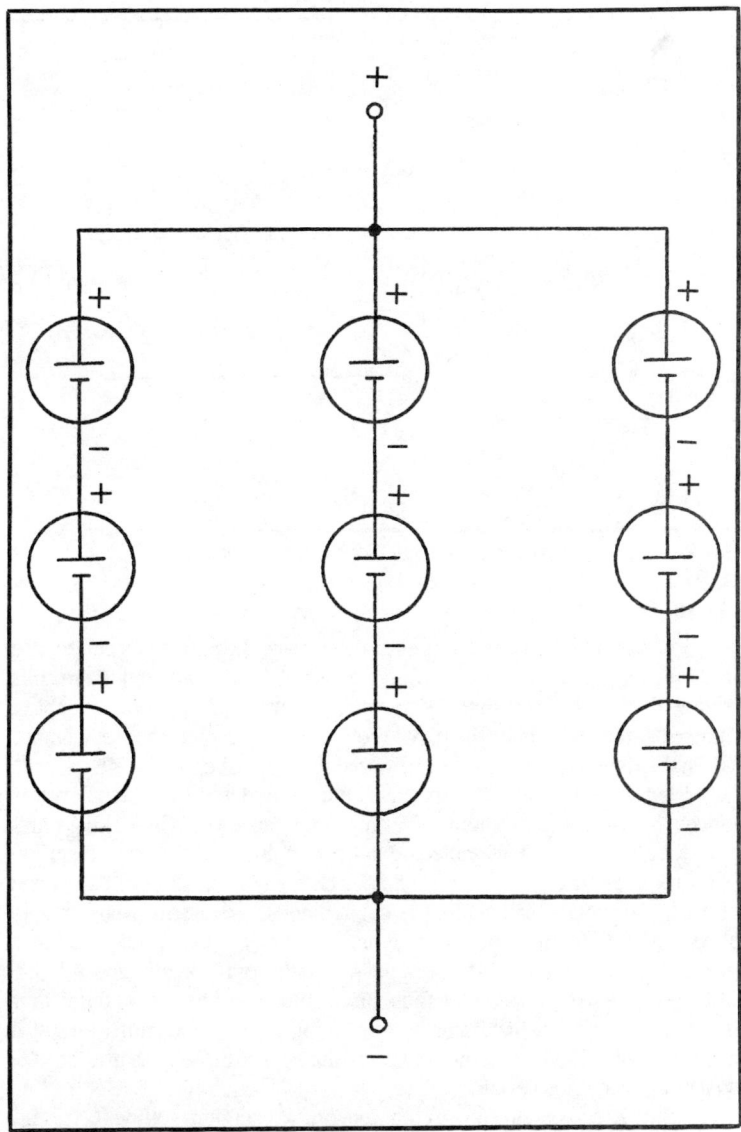

Fig. 2-8. Series-parallel solar cell connection.

series to the tops of the three series blocks. If the supply required 1.35 volts at a higher current level of, let's say, 0.4 amperes, then another string of three series cells would be added in parallel. Figure 2-9 shows several different solar cells arrangements using the basic cell in this discussion. Note the different output combinations which may be arrived at by adding cells.

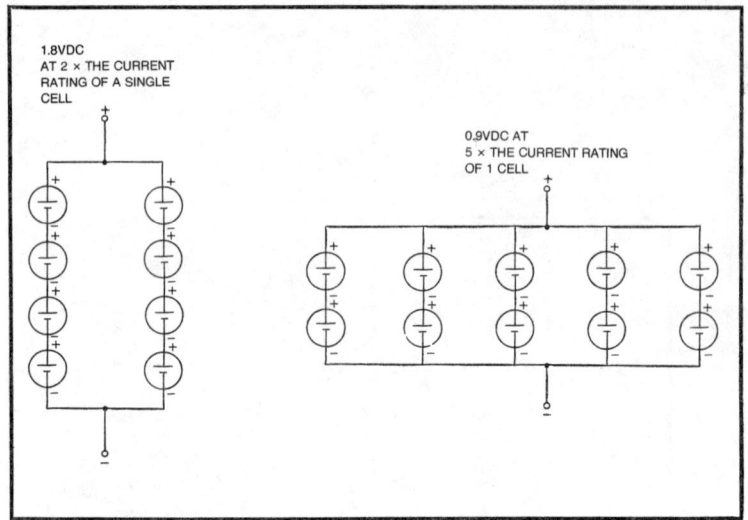

Fig. 2-9. Other types of wiring configurations for solar cells.

PHOTOCONDUCTIVE CELLS

Photoconductive cells, or photoresistors as they are often called, are passive photoelectric devices. Unlike the solar cell, photoconductive cells change their internal resistance to current flow based upon the amount of light present. The higher the light level, the lower the internal resistance. When the light begins to dim, the internal resistance begins to rise.

Photoconductive cells are available with different internal resistances, but all of them exhibit a very high resistance when in darkness and a low resistance when subjected to bright light. A cadmium sulfide photoresistor available from Radio Shack sells for less than $1.25 and offers a wide resistance range. In total darkness, the internal resistance is about 5,000,000 ohms, or 5 megohms; but in bright light, the resistance drops to only 100 ohms. It offers peak sensitivity to yellow-green light. Most photoresistors will not handle large amounts of power, and this unit is no exception. Its maximum power rating is only 200 milliwatts at a maximum of 170 volts. This power rating is a little less than that of a quarter-watt carbon resistor.

Allied Electronics offers a complete line of cadmium sulfide (CdS) and cadmium selenium (CdSe) devices which are designed to be most sensitive to varying light frequencies, depending upon the device purchased. Some will closely match the visual response of the human eye, while others are designed to be responsive to frequencies which lie out of our visual ranges. Some of these have power ratings of two watts but require finned heat sinks to dissipate this power. Without the external sinks, the components are rated at one-half watt. Prices are still in the low category, especially when compared to solar cells. In the Allied Electronics 1980 catalog, the highest priced photocell was only $3.20, and this was for a 2-watt unit.

It is interesting to note the resistance ranges. Some have ratios of 1:100 between light and dark, while others have ratios of 1:10,000. This means that the resistance in total darkness is 100 or 10,000 times the resistance when a light of 2 footcandles is played upon the surface of each device, respectively. Maximum resistance exhibited by some devices was on the order of one billion ohms, while minimum resistance in others might be less than 100 ohms. Maximum voltage ratings were around the 300 volt mark, indicating that photoresistors will not find a lot of application in high-voltage work, although series connections would improve this.

Just as solar cells may be wired in the same manner as dry-cell batteries, CdS photocells may be wired just like resistors in an electronic circuit. The properties displayed by conventional resistors will also be displayed by these cells, and many of the electronic calculations which apply to resistors will also apply to CdS cells.

In series circuits, resistance combines. Ten resistors with a value of 1 ohm each will exhibit a total resistance of 10 ohms when combined in the series circuit shown in Fig. 2-10. Power ratings also add in series or parallel circuits with each unit, so if the value of each resistor is 1 watt, the total power-handling capability of the series string will be 10 watts.

By substituting photoresistors for the standard resistors in the previous schematic, we arrive at the schematic shown in Fig. 2-11. The schematic symbol for a photoresistor is very similar to that for a standard resistor. Assuming that the value of each photoresistor is 100 ohms under bright light, a series connection of 10 of them as shown would exhibit a total resistance of 1000 ohms (if all 10 photoresistors were subjected to the same intensity of bright light simultaneously). If the power rating of a single photoresistor is 250 milliwatts (one-quarter of a watt), then 10 units connected in series would offer a total power capability of 2.5 watts. If the resistance of one cell in absolute darkness is 5 megohms, then the total circuit will have a resistance of 10 times this amount or 50 million ohms.

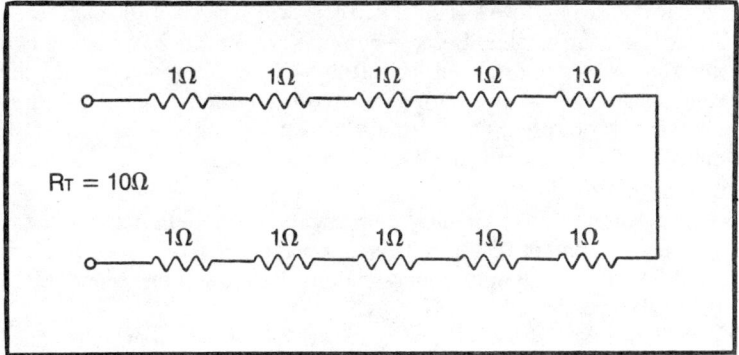

Fig. 2-10. Series connection of carbon resistors.

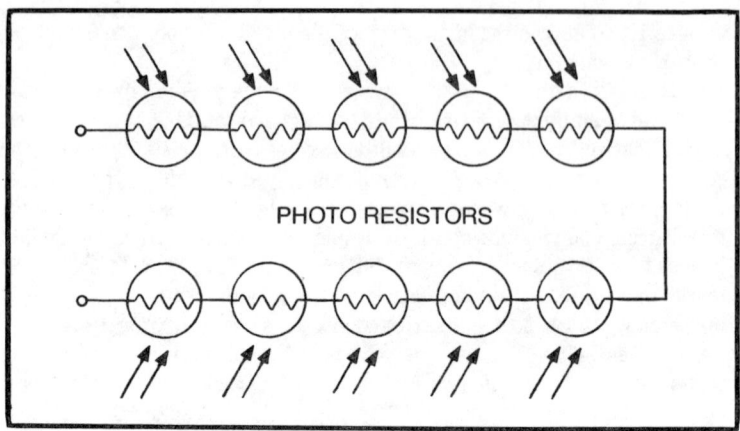

Fig. 2-11. Series connection of photoresistors.

Resistors in parallel divide resistance. Figure 2-12 shows a parallel circuit composed of two photoresistors of equal values. The formula for determining two overall parallel resistance to current flow is

$$\frac{R1 \times R2}{R1 + R2} = R_t$$

This looks complicated but simply means that the total resistance (R_t) is equal to the resistance of one device times the resistance of the other divided by the sum of the two resistances. Assume that the photoresistors in the schematic each have a resistance of 100 ohms. Working these values into the formula, we would come up with

$$\frac{100 \times 100}{100 + 100} = R_t \text{ or } \frac{10,000}{200} = 50.$$ Fifty ohms is the answer.

It is not necessary to go through this formula every time you combine photoresistors or, for that matter, standard carbon resistors. When photoresistors of EQUAL VALUES are used in a parallel circuit, we can say (for all practical purposes) that the resistances divide by the number of cells used. In the above example, two cells with a value of 100 ohms each are combined in parallel. Because two cells are combined, we divide two into the value of a *single* cell, or 100 ohms. The answer is 50 ohms, the same answer as derived from the more complex formula. If three photoresistors were used, we would divide the value of a single cell (100 ohms) by the number of cells (3) and arrive at a total resistance value of about 33 ohms.

Just as in series circuits, power adds in parallel circuits; so two quarter-watt photoresistor cells would present a total power-handling capability of one-half watt. Simply multiply the power rating of one unit by the total number of units used in the circuit.

Figure 2-13 shows a combination series-parallel circuit comprised solely of photoresistors. Figuring total circuit resistance here is most difficult, although the power-handling capability is easily enough arrived at by multiplying the value of one cell by the total number used throughout.

To accurately figure the total resistance, we must first break the circuit down into separate sections. Figure 2-14 shows how this is done. Since each series string contains four 100-ohm units, we know the total resistance of this "leg" of the circuit is 400 ohms. Instead of seeing this leg as four photoresistors, it would be simpler to view it as one photoresistor with a value of 400 ohms. The same would be true of the three other matching legs. So, what we really have is four 400-ohm resistors connected in parallel. Going to the formula for resistance in parallel, we would divide the value of one resistor leg (400 ohms) by the number of legs (4). The total circuit resistance is 100 ohms, the same resistance value as for a single photoresistor.

You may be asking yourself at this point, what is the purpose of going through this complicated wiring process to arrive at a circuit resistance which is equivalent to a single unit? Wouldn't it be simpler just to use a single photoresistor? The answer is certainly yes—if you don't require the other ratings this complex circuit provides; namely, increased power-handling capabilities. Let's assume you require a 100-ohm value photoresistor for connection to an electronic circuit, but this device must have a power-handling capability of at least three watts. No single unit made within a reasonable price range is available, so we must make up a circuit

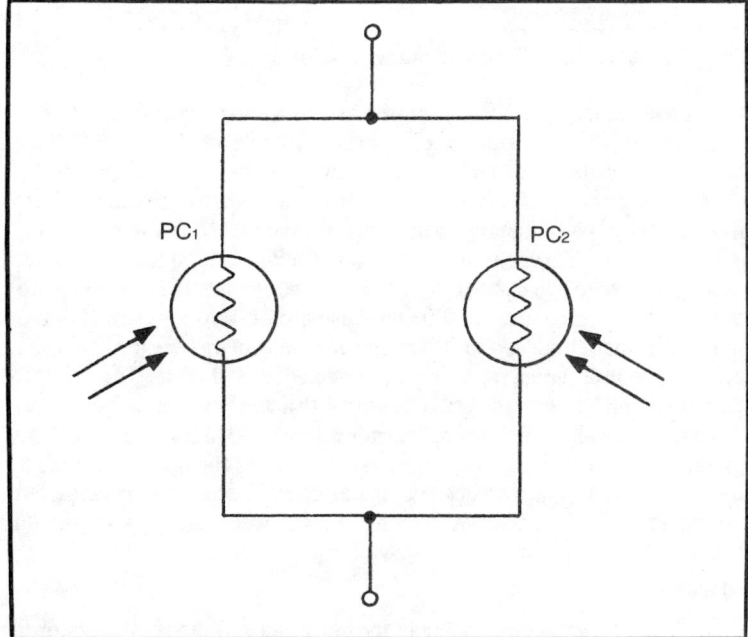

Fig. 2-12. Parallel circuit composed of two photoresistors.

29

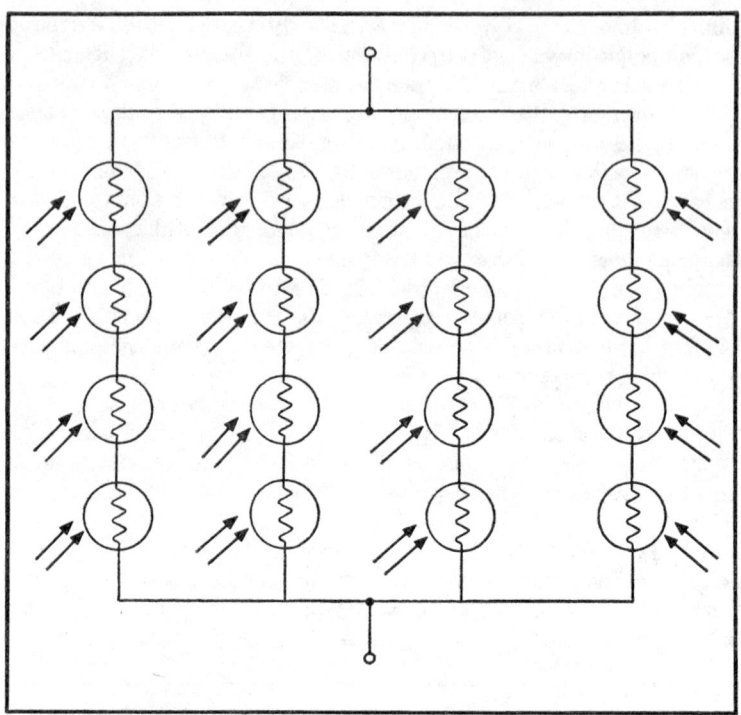

Fig. 2-13. Series-parallel circuit formed with photoresistors.

which uses at least 12 photoresistors, assuming that each is rated at 250 milliwatts (0.25 watt). However, we must use a total of 16 photoresistors, because 12 of them connected in this same series-parallel circuit would provide a resistance of only 75 ohms. (This representative circuit is shown in Fig. 2-14.) We could take away one of the parallel connections and add another unit to each series leg, as is shown in Fig. 2-15; but now the total resistance would equal about 133 ohms. So, we see that there is no way to obtain a total resistance of 100 ohms using only 12 units, which is the minimum number needed to obtain a power rating of three watts. We can't go lower in unit numbers, so we must add additional units. The original circuit exhibits a resistance of 100 ohms with a power rating of four watts, assuming individual units of one-quarter watt each. This is in excess of the minimum power-handling capability required but is the minimum number needed to arrive at the correct resistance figure. The excess power is even desirable, because each device tends to run well below its maximum ratings and may provide a longer service life.

TRANSISTORS

To explore the functioning of the solar cell and the photoresistor in electronic circuits and to continue farther in the discussion of more sophisticated light-dependent devices, it is necessary to talk about a very common electronic component which is the one most often used in all electronic circuits today.

The transistor shown in Fig. 2-16 performs a vast amount of electronic functions. When a small signal is applied between the emitter and the base junction, the transistor begins to conduct current through the collector-emitter circuit.

A transistor is similar to two diodes connected at their cathodes. The cathode junction is the base, while the anode terminals serve as the emitter and the collector. This applies only to a *bipolar transistor*. There are two types of bipolar transistors in common use, the npn and the pnp. Both are shown schematically in Fig. 2-17. Transistors are used for switching purposes, for amplification of power, and for many other electronic applications. The designations of bipolar transistors have to do with the types of semiconductor materials used to make them, and more accurately, in the ways this material is layered or "sandwiched" to form the finished transistor.

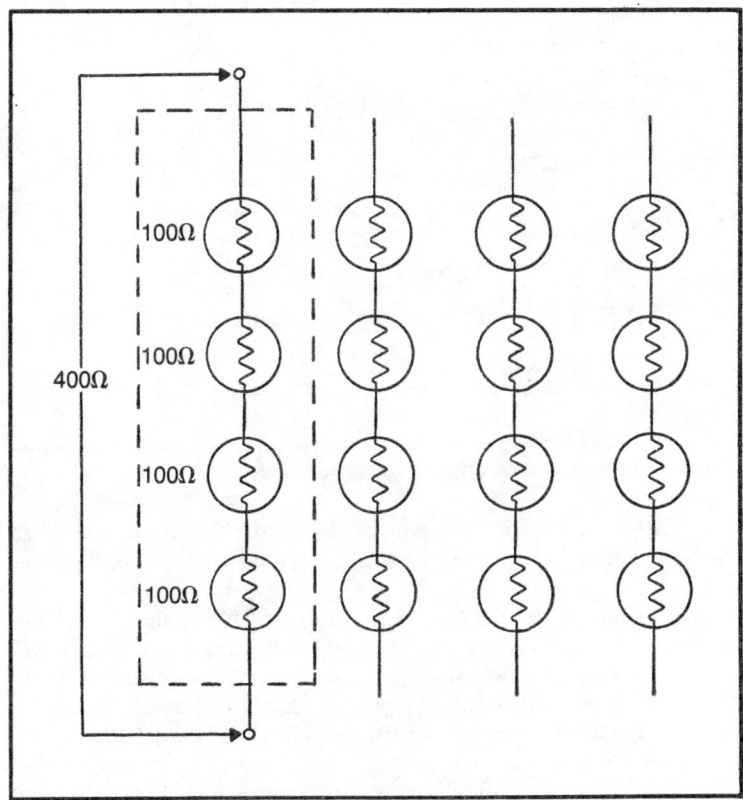

Fig. 2-14. Complex combination of photoresistors.

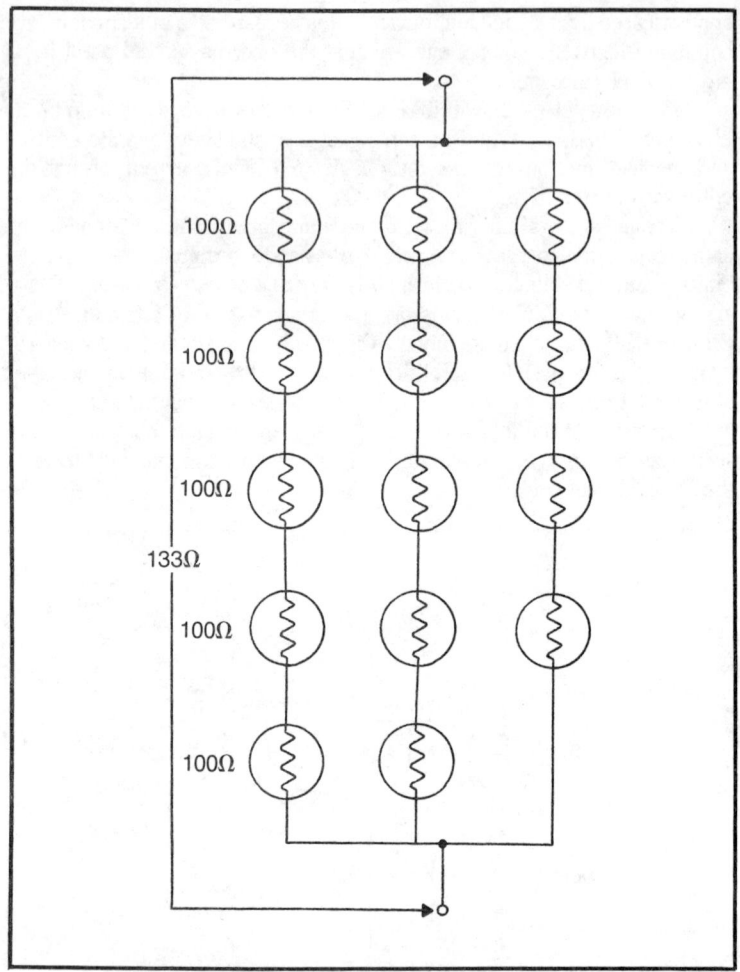

Fig. 2-15. Modified version of former complex circuit.

The conductivity of a transistor is controlled at its base-emitter junction. A small-value varying signal at this junction will result in the transistor conducting a large amount of current through its collector-emitter circuit, but the variation at the base is still maintained through the collector and emitter. The on-off cycle of the signal at the base will produce an equivalent on-off cycle in the conduction of higher current to the collector-emitter. This could also produce an opposite reaction of off-on at the collector-emitter for an on-off action at the base.

Figure 2-18 shows a transistor circuit of very simple proportions. The base resistance controls the current flow between base and emitter and sets up the conduction value of the transistor. If this base resistor value is

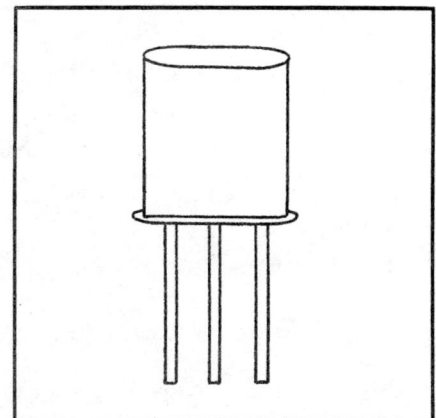

Fig. 2-16. Common transistor.

lowered, more current will be conducted and the amount of conduction through the collector-emitter will change.

Figure 2-19 shows how a photoresistor may be substituted for the base resistor. Now, the amount of current flowing in the base leg of this circuit will be directly dependent upon the amount of light striking the treated surface of the photocell. When there is almost no light, the resistance is very high and little current will flow; but as the light intensity increases, the resistance of the cell decreases, current flows in the base leg, and conduction occurs between the collector and emitter.

The value of a photoresistor varies over such a wide range, so the previous circuit is not nearly as practical as the one shown in Fig. 2-20. Here, a base resistor is installed in series with a photoresistor. Both of these devices are in series with the base circuit leg. The fixed resistor determines the minimum amount of resistance of the circuit. If the value of the fixed resistor is 1000 ohms and the minimum value of the photoresistor is 500 ohms, then the minimum resistance of the base leg circuit will be the sum of these two values (resistances in series add), or 1500 ohms. The fixed resistor is chosen to complement the specific transistor, circuit, and photoresistor used. An alternate method may be utilized in some circuit configurations where the photoresistor is wired in parallel with a fixed base resistor. This circuit is shown in Fig. 2-21. This is a good combination when the minimum value of the photoresistor is still too high.

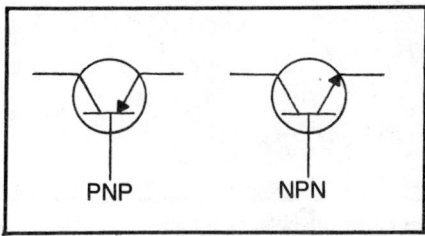

Fig. 2-17. Schematic symbols of bipolar transistors.

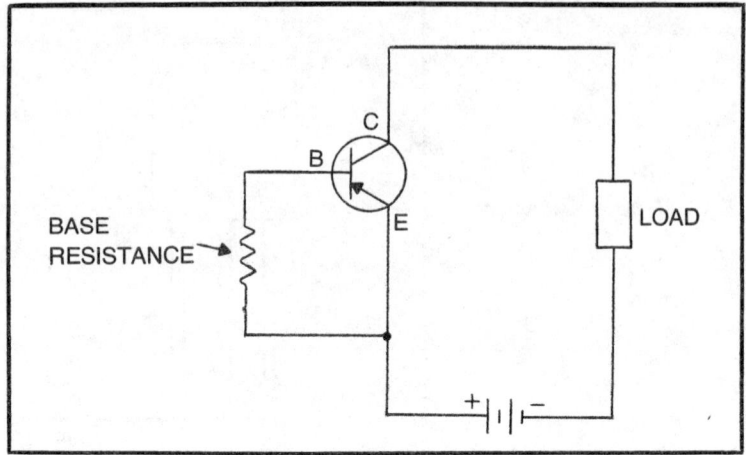

Fig. 2-18. Simple transistor circuit.

If the fixed resistor is of the same value as the minimum resistance of the photoresistor, the total resistance in the base leg circuit will be half of the minimum resistance of the cell when this unit is subjected to intense light. This, then, would be the minimum base circuit resistance.

In on-off operations, a photovoltaic cell may be put to good use in the base leg of a similar transistor circuit. The schematic shown in Fig. 2-22 uses a npn transistor with a single solar cell in the base leg. As long as the cell is not exposed to intense light, the collector-emitter junction will not conduct current. This is due to the fact that no current is flowing at the base-emitter junction. But when light strikes the cell, it produces current which activates the circuit. Current flows to the collector-emitter junction

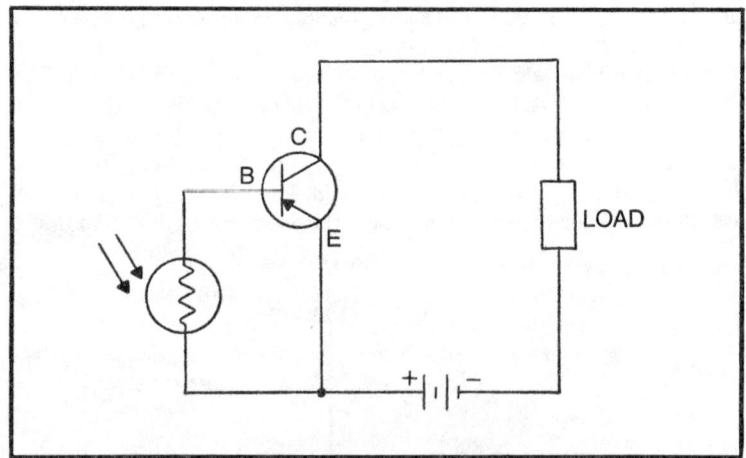

Fig. 2-19. Use of photoresistor in base circuit of bipolar transistor.

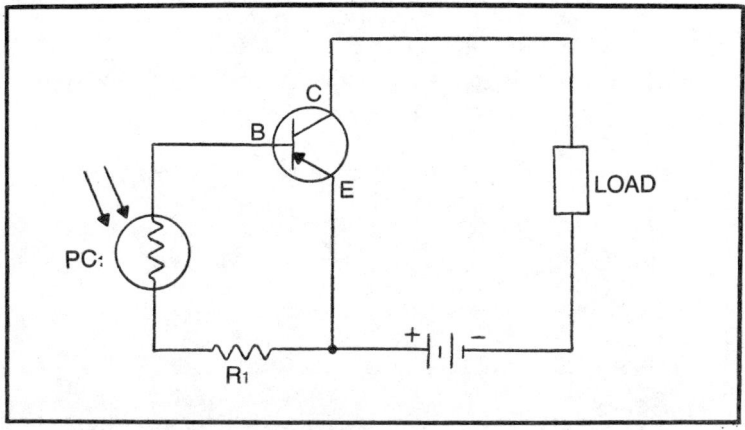

Fig. 2-20. Photoresistor-resistor connection in base leg of bipolar transistor.

and, thus, through the load. The load in this case might be the winding of a relay to which a simple incandescent light circuit has been attached. When the relay is activated, its contacts close and send operating current to a light bulb. This circuit is shown in Fig. 2-23. This is a common circuit used for activation of devices by sunlight. To activate a circuit at the absence of sunlight simply requires the placement of the normally open relay specified with one which is normally closed. This latter device will keep the lighting circuit activated because its contacts are always closed when no power is being conducted through the relay coil. At night when no light is striking the surface of the solar cell, the transistor does not conduct and current does not flow through this coil. The relay is not activated, so the light circuit it controls stays on at night. But when the sun comes up the next morning, light rays strike the solar cell, causing the transistor to conduct current through the relay coil. The relay is activated and the

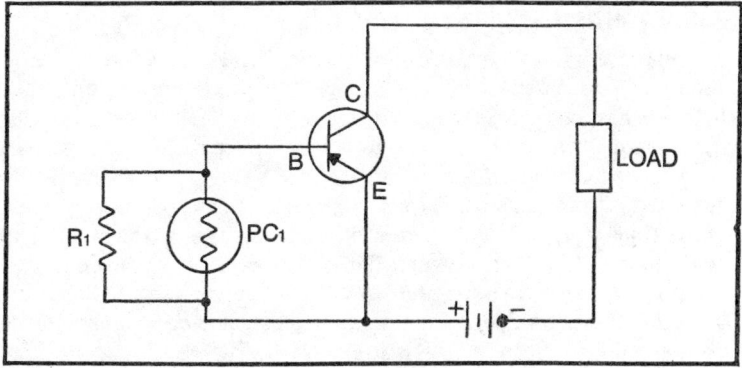

Fig. 2-21. Parallel connection of photoresistor and resistor in transistor base circuit.

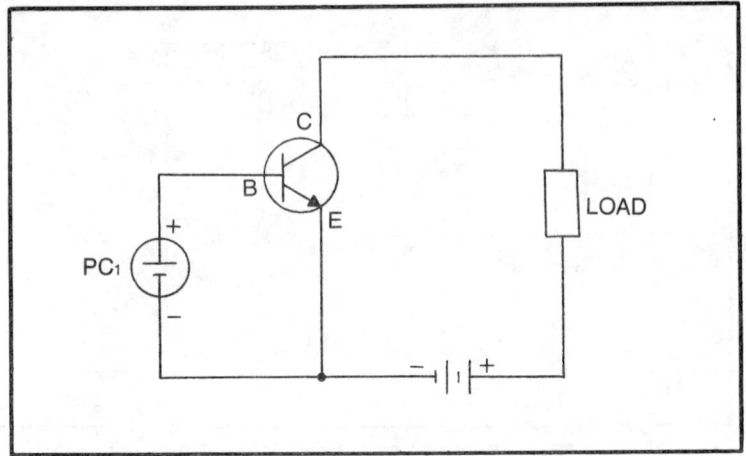

Fig. 2-22. Solar cell inserted in base leg of n-p-n transistor.

normally closed contacts completing the power circuit to the light suddenly open, breaking the lighting circuit. The result is the light goes off when the sun comes up. This is the basic circuit used in most day/night switches.

There are many more variations of electronic circuits using the solar cell or the photoresistor in place of, or to augment, standard fixed components. Any circuit which can be controlled by a variable resistance can be controlled by a photoresistor. This means that light intensity can ultimately control the circuit. Any circuit which can be controlled by a small electrical current may be controlled by light intensity through use of a solar cell. Many people think of solar projects as being those which are powered by light. This is not always true, as many projects will have conventional power sources but will be controlled by light variation.

PHOTOTRANSISTORS

Photocells are composed of a junction made when two types of semiconductor materials are pressed together in such a manner that when light strikes near a junction, an electron flow is released. If this transparent junction is backed up by another semiconductor layer called a collector, photoconductive current generated by the photocell junction is amplified as in a transistor. The sensitized area becomes the base junction of the transistor; and when light strikes this area, it produces an input signal to the device. This is very similar to a circuit which can be formed using a photocell and a p-n-p transistor. This circuit is shown in Fig. 2-24. The phototransistor does away with the separate photocell or photodiode portion of the circuit and houses it in one unit.

Since a phototransistor is really a combination of the circuits previously discussed in one component housing, the reader might wonder

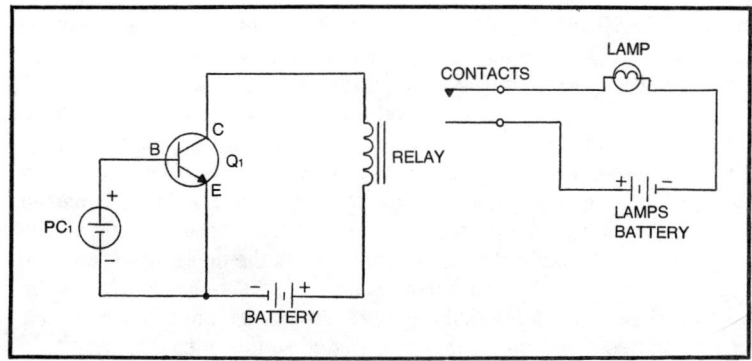

Fig. 2-23. Light-activated switching circuit using solar cell, transistor, and relay.

why circuits using discrete components (transistor, resistor, photocell) are still used. The main reason is that more versatility may be had by converting present circuits using standard transistors rather than by designing new circuits around the limited number of phototransistors available. Then too, from a hobby standpoint, it is usually easier to obtain common transistors and inexpensive photocells than it is to buy phototransistors. Radio Shack offers only one phototransistor, which sells for less than one dollar. This is far less expensive than buying a standard transistor and a photocell, but this is a single transistor type. It will not work in every solid-state circuit, only in those which are designed around the operating parameters of this device. Sure, if you live in an area where there are abundant sources of sophisticated electronic parts, you may be able to get a phototransistor of the exact type needed. Chances are, however, that you do not fit into this category and will have to make do with the limited

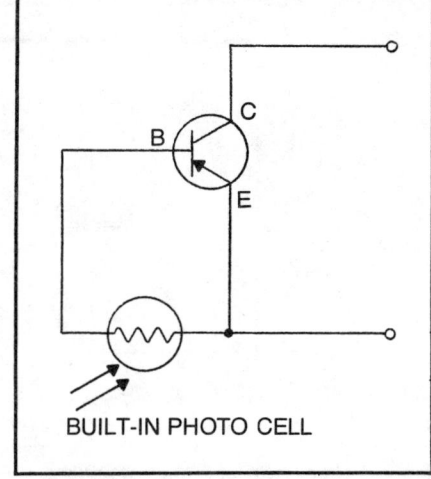

Fig. 2-24. Electronic equivalent of a phototransistor.

resources available. In this day and age, it is a relatively simple task to locate a common transistor and small photocell. It is much more difficult to obtain a specific phototransistor within a short period of time.

Phototransistors are excellent devices for certain applications. On-off control is one of these when the intensity of the light source can be carefully maintained. But this is not true when one must depend upon sunlight to activate or deactivate a circuit. A previous circuit is shown again in Fig. 2-25. This is the day-night switch discussed earlier, but an additional component has been placed between the photovoltaic cell and the transistor base. This is a variable control or potentiometer wired in the variable resistor mode. With this control, the current flow through the cell can be limited. The control may be adjusted so that adequate current to cause the transistor to conduct will flow only when the cell is producing a certain amount of current. This would be directly related to the amount of sunlight striking the cell. For instance, if your night lights were being turned off too early because the early morning light produced sufficient current flow through the cell, the variable resistor could be set to increase the resistance to this current flow. This would mean that the transistor would not begin conducting until more current was produced by the base circuit. The lights would not be turned off until the sun became more intense as the morning drew on. This type of control would not be possible using a phototransistor because the circuitry between the photocell portion and the transistor base is contained on a single crystalline chip and housed in one container.

If the opposite condition were occurring and the lights being turned on too early in the evening, the circuit shown in Fig. 2-26 might be used. Here, a photoresistor is placed in the base leg with a variable control in series with this circuit. This control could be adjusted to a point where enough current would flow to cause conduction under the desired lighting conditions. A photoresistor would have to be chosen which also corres-

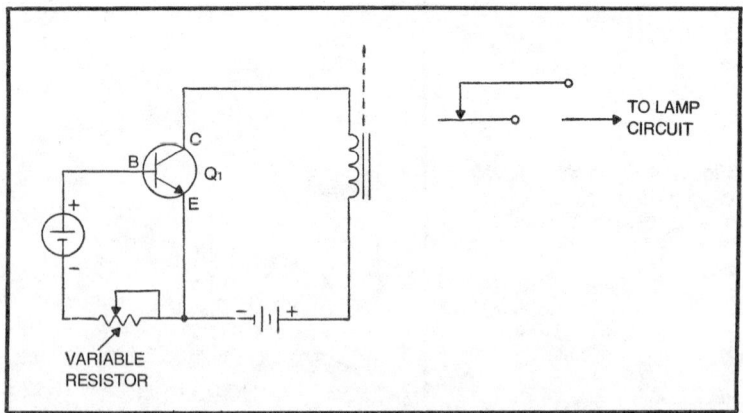

Fig. 2-25. Light-activated switch with variable resistor control.

ponded to the resistance needed in the circuit at the general light level desired. Is the resistance of the cell were too high, the circuit might be altered as is shown in Fig. 2-27. Here, an additional variable control is wired in parallel with the photoresistor. This directly affects the internal resistance of the cell, as seen by the rest of the circuit, and allows for an even greater control of triggering sensitivity and adjustment.

This discussion on transistors, phototransistors, and photocell devices has attempted to demonstrate the many circuit variations that can be had when photosensitive devices are combined with the more standardized components used in electronic circuits today. Many of the ideas presented in this chapter will be directly reflected in the construction projects later on. These principles may be carried into your future experimental circuits, modified and improved upon considerably. The circuits presented here are very simple, single-stage devices. These may be used as shown or may be called upon to provide driving current for more complex electronic stages. Sometimes, even combination circuits consisting of photovoltaics and photoresistors will be used in the base-emitter legs of some transistor circuits.

OTHER LIGHT-SENSITIVE DEVICES

The major devices which will be included in the forthcoming projects have been discussed in this chapter. However, there are many other devices which are photoelectric in nature that may be encountered from time to time. One of these is the LASCR (light-activated silicon controlled rectifier). This device allows for the switching on and off and variable control of very high currents through light intensity. You may also encounter in your experiments optical oscillators, photochopper cells, and

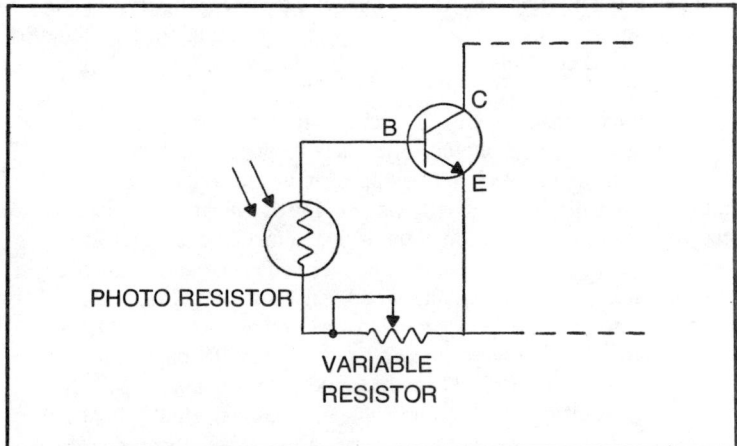

Fig. 2-26. Another light-activated switch using a photoresistor and variable control.

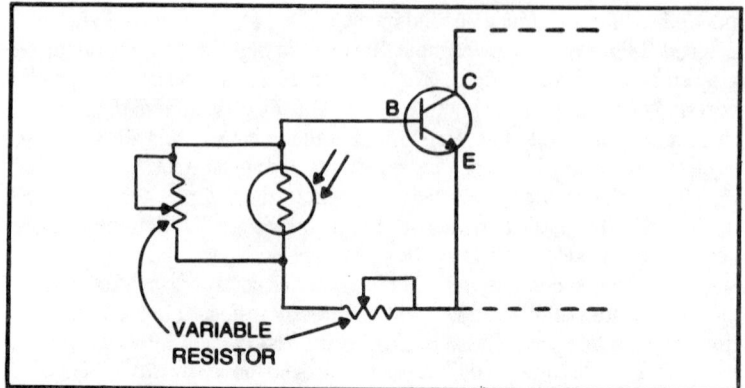

Fig. 2-27. Photoresistor connected in parallel with variable control.

others. These all work around the same effects which the devices discussed are based upon.

SUMMARY

The variations which may be had in the control of electronic circuits through light intensity and frequency are arrived at through non-complex application of common electronic components. It is interesting to note that all of the photocell devices discussed are merely modifications of standardized components which have been with us for many years. Some are solid-state, light-controlled equivalents of these components. The photoresistor, for example, is a light-controlled counterpart of a carbon resistor. The phototransistor is a light-controlled transistor, whereas the solar cell is the light-powered equivalent of the common battery. While all of these comparisons may not be completely accurate in the scientific sense, they will certainly suffice from a building and experimenting point of view.

The function of any solid-state component can probably be duplicated by its equivalent in a photoelectric cell device. The light-controlled equivalent of a standard circuit will perform the same functions, only the method of control will have been changed. The day-night switch could have accomplished the same function by having someone manually activate a wall switch. The circuit presented accomplished the same function. In both instances, the light would have been turned on. Using the light-controlled method, the desired action was pre-controlled by the circuit which reacted to the rising and setting of the sun. Through electronics, circuits are devised which allows humans to determine control actions. These actions may be in the form of sound, current flow, temperature change, or, in this case, light intensity. All of these conditions can be thought of as actions; it is up to the experimenter to build the circuits which respond to these actions with the desired reactions.

Chapter 3
Electronic Project Building

Having discussed the various components which will be used to build the circuits in this text, it is now necessary to delve heavily into the mechanics of actually putting together an electronic project. If you have some experience in electronic kit building, this will be most helpful. If you are new to electronics, the information contained in this chapter should teach you enough to begin building some of the simpler projects. After two or three projects are completed, more difficult circuits may be attempted with the confidence of adequate technique.

Most of the projects presented in this book can be best constructed on perforated circuit board material which is available from hobby stores and through mail order outlets. Sometimes called "perf" board, this material consists of a dielectric or insulating strip which is fairly rigid. Through this material, many rows of holes are punched to allow for the insertion of various component leads and hookup wires. Figure 3-1 shows a sample piece of perforated circuit board which is available in sizes measuring 1" square to larger pieces measuring 6" by 10". Some types will have wide-spaced holes while others will be more closely spaced and numerous. Thickness of the board will determine its weight and flexibility. The greater thicknesses are desirable when these boards must support a large number of medium-sized components, such as electrolytic capacitors or audio transformers. The thin sections are best used for miniature or lightweight components and where weight may be a consideration. Weight is not important when building the projects outlined in this book. Most circuit boards weigh less than one ounce, regardless of the material thickness.

Perforated circuit board is ideal for the experimenter or hobbyist because it provides a great deal of versatility in building many different

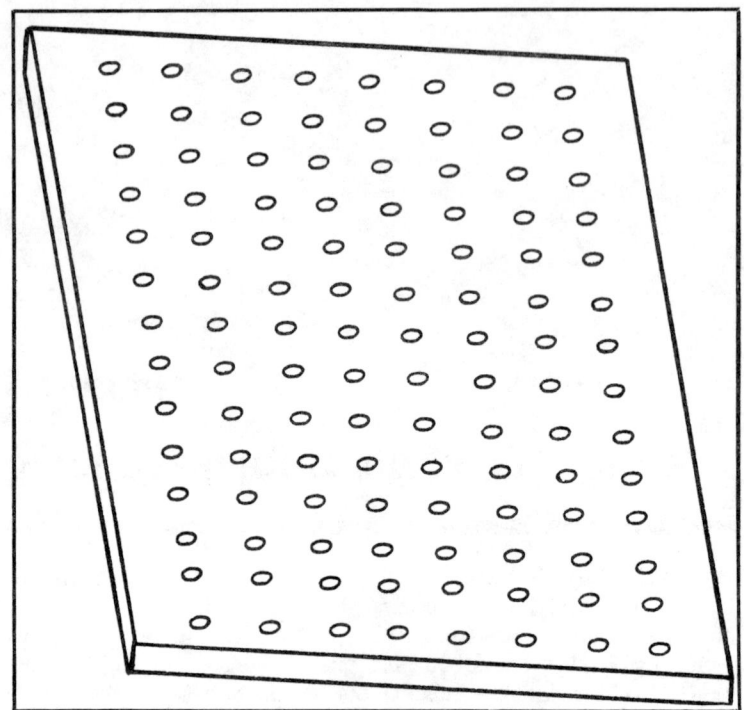

Fig. 3-1. Section of perforated circuit board used for electronic building.

projects. It is quite inexpensive, with even the larger sections selling for less than $3. Small sections of perf board may be purchased in quantities of five or more, usually for less than a dollar. Check around, as many hobby stores may sell perforated circuit board in quantity packs at surplus prices.

For the more experienced builders who desire the convenience of printed circuit board construction, it is possible to design these boards at home using printed circuit board kits. These consist of circuit boards made from a phenolic material and completely covered on one side with an ultra-thin copper coating. Figure 3-2 shows this type of circuit board. The copper sheet is used rather than hookup wire, to connect the various components together in the circuit. Only very small strips of the copper coating will be left on the finished product. The rest of it will have been etched away in the construction process.

First of all, the circuit schematic is examined and a component placement configuration is established. Figure 3-3 shows a simple electronic circuit in schematic form which we will use as the basis for designing a printed circuit board. From this schematic, we can determine the number of components which will be required and the manner in which they are to be connected. Figure 3-4 shows a drawing of how the components would be physically mounted on the circuit board.

Fig. 3-2. Printed circuit board material.

Looking at the board from the other side, we now sketch the connections which need to be made from the various components to other parts of the circuit. The components are indicated by the dotted lines. This drawing should be the same size as the circuit board.

Notice that a large dot is drawn where a component lead pierces the circuit board and that several connections may reach a junction at one of the larger oval dots. All of this must be planned in advance, because once the circuit board is completed, it will be very difficult to make changes, especially those which require the addition of more conductors. Once you are certain that your drawing is correct by comparing it with the schematic, it is time to transfer this drawing to the bottom side of the circuit board (the portion containing the copper coating). This may be done

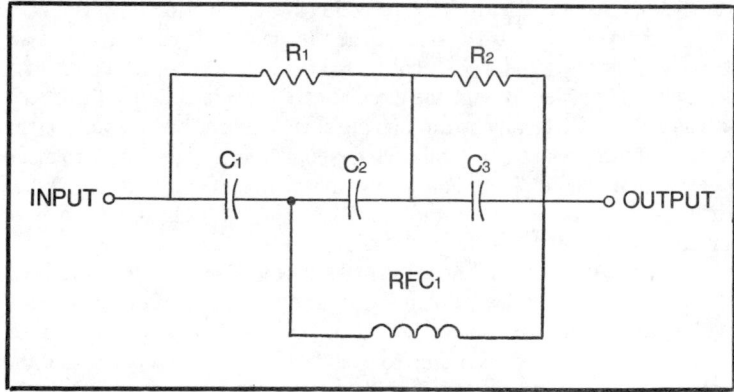

Fig. 3-3. Simple electronic circuit to be built on printed circuit board.

43

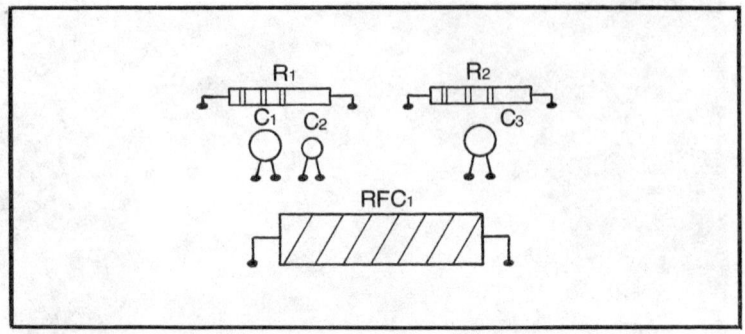

Fig. 3-4. Component placement on printed circuit board.

by cutting out the inked-in portions of your drawing or by simply retracing it as closely as possible onto the copper material.

Resist-ink is used to make the drawing on the copper-clad circuit board. This ink usually comes in a special marking pen which may be purchased from most hobby stores for less than two dollars. The ink is a special type which will resist the etchant solution which will be used later to remove unwanted copper from the board. Make certain that the lines you draw are continuous. If the ink misses a spot on the copper which is intended to be saved, this area would be etched away during later processes. The copper will remain only where it is covered by the resist-ink. Make certain that no ink has been allowed to drip onto a portion of the copper material which is to be etched away. Most kits contain a solvent with which you may remove the resist-ink when errors have been made. Figure 3-5 shows a circuit board which has been properly prepared for the etching process.

Also included with most printed circuit board kits is a small drill bit with which to make the desired holes for insertion of component leads. These holes must be drilled before the etching process begins to prevent small sections of copper from being pulled away. If a drill bit has not been included in your kit, a 1/16" size will be adequate. This may be purchased from most hardware stores. A hole is to be drilled wherever a component lead is to be pushed through the circuit board for connection to the circuit wiring. This will usually be through the small dots and ovals placed on the board with the resist-ink. Drill from the copper side of the board to make certain that a hole is placed through the center of each ink dot. When drilling is complete, insert the resist-ink pen into each of the holes to protect them from the etching solution.

Examine your circuit board to make certain that all holes have been drilled and that the drawn circuit is absolutely correct. After this stage it will be too late to make changes. If all seems proper, the etching solution may now be prepared according to manufacturer's directions. Often, the solution may be used directly from the bottle, but occasionally it must be diluted. Once prepared, this solution is poured into a plastic tray usually supplied with the kit. This tray must be large enough to accommodate the circuit board but not so large as to spread the etchant out to a point where it will not completely cover the board.

Plunge the board into the etchant solution, making certain that the entire copper surface is completely covered. It is a good idea to agitate the solution by gently shaking the tray to get rid of all air bubbles which may build up between the copper surface and the etchant. Total time of immersion will depend upon the strength and the amount of the solution, and the size of the circuit board. Warning: This solution is toxic and should not be breathed nor allowed to come in contact with the skin.

Once the manufacturer's stated immersion time is up, immediately remove the circuit board from the solution and allow it to dry. Do not leave the board in longer than the stated time, as the resist ink may not be capable of protecting the copper area it covered much past this point. When the circuit board is dry, wipe away all foreign material and examine the finished product. Look closely to make certain there are no breaks in the copper strips which remain. The resist-ink solvent may be used to rid the copper strips of their black covering. This will allow you to better examine the copper strips which interconnect the various electronic components. Look, too, for signs of stray copper material which has not been completely etched away. This may be removed by using a miniature screwdriver in slow, scraping motions.

If you notice that one of the copper strips has been broken by the etching away of a small section, this may be easily repaired with a soldering iron and a small piece of stranded hookup wire. Figure 3-6 shows a break in a printed circuit conductor strip. This is repaired by carefully cleaning the areas on either side of the break of all foreign matter, such as resist-ink, oil, or dirt. This may be done by lightly scraping the surfaces with a small screwdriver. Do not scrape so hard that the copper material is removed from the board. When the areas on either side of the break are shiny, tin them with a small amount of solder and soldering iron. Now, place the strands of uninsulated hookup wire across the break and apply

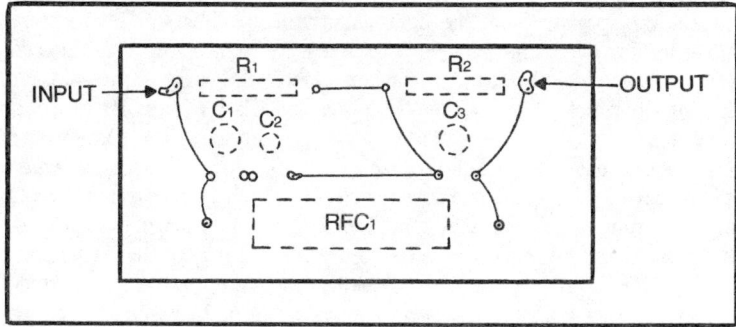

Fig. 3-5. Printed circuit board material which has been prepared for the etching process.

heat once more until the gap is completely bridged. Check the circuit continuity with an ohmmeter to make certain the repair is of good quality.

If all has gone well in your printed circuit board designing and etching process, the remaining copper material on the bottom of the board should match the drawing which was originally made on paper. If it does not, examine the two closely to determine just what is in error and correct the error, if possible. This will only be practical when excess copper has been left on the circuit board. If too many sections have been etched away, it will be necessary to start over again or to make up the missing sections with insulated hookup wire.

While printed circuit boards are easier to wire than are their perforated counterparts, the construction of the former may far outweigh its disadvantages. Printed circuit boards provide a much more stable base for components due to the fact that everything is tied down, even the connections between components which take the form of loose hookup wiring in other types of circuits.

Projects which control frequency or other critical parameters may not operate properly or may "drift" if components are allowed to move even a little. Due to the highly stabilized manner of building using printed circuit boards, it may be desirable to use this technique where critical conditions require high component stability. It should be pointed out that all of the circuits in this book will work as designed when built on perforated circuit board, providing that proper construction technique is used. Printed circuit board requires the same attention to detail and cannot be used to make up for poor building practices.

If the reader desires to make up printed circuit boards for the following projects, he or she is encouraged to do so because of the increased experience which will be gained. If the reader would rather stay with the more customary perforated circuit board technique, this should make no difference whatsoever to the successful operation of the project attempted.

In building the projects presented in this text, the reader will be working with transistors, diodes, integrated circuits, resistors, and capacitors, in addition to the light-sensitive components. When putting together any electronic project, it is necessary to have certain skills and tools to complete the job in a manner which will assure successful and lasting operation. The tools will be dealt with a bit later. At present, neatness may account for quite a lot. A neat circuit is one where component leads do not tend to come in contact with one another when they are not supposed to. Sure, a circuit which is rather hastily put together may function normally, but it has a greater chance of failing, if not at first, at some later date. When circuits are wired in a haphazard manner, the builder often loses touch with which component leads go to which circuit contact point. The builder, more or less, loses control of the situation. It is necessary to arrive at a game plan for the construction before the actual building is started. For this reason, each of the projects

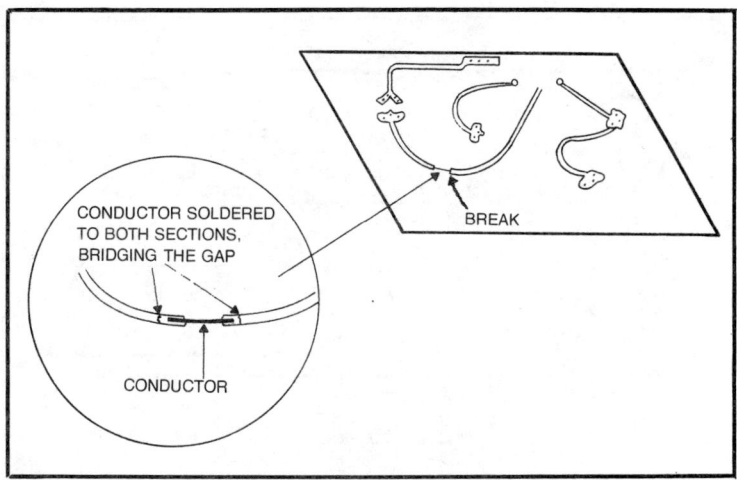

Fig. 3-6. A break in a printed circuit board conductor strip.

will contain a suggested parts layout which does not have to be followed exactly but will lend a hand in deciding the placement positions of basic circuit components. By sticking with this plan, a form of continuity is established from the start and will help the builder to carry through to the end of the project.

It is not the author's aim to heavily restrict those persons desiring to build some of the many projects contained in this text. As a matter of fact, the reader is encouraged to experiment to any extent he or she desires; but, experiment along established lines. Continuity is still required when deviating from a circuit diagram or from a suggested component placement form.

Regarding component placement, it is left up to the reader to determine the manner in which the components are to be installed on the circuit board. There are two established ways of mounting components. Both of these methods involve pushing component leads through holes in the perforated board and making all solder connections on the opposite side of the board. The most standard of these two methods is *horizontal mounting,* which is shown in Fig. 3-7. Here, components such as resistors and capacitors are placed lengthwise along the circuit board with component leads pushed through the holes which are spaced approximately the component's physical length from each other. This method is suggested for those who are relatively new to electronic project building. Once the component leads are placed through the circuit board, they are wrapped with other component leads and hookup wiring before being soldered into place. The horizontal mounting method tends to spread components out. This allows for a simpler viewing of the circuit sections but also means that the completed project will fill a larger physical space than when the second method is used.

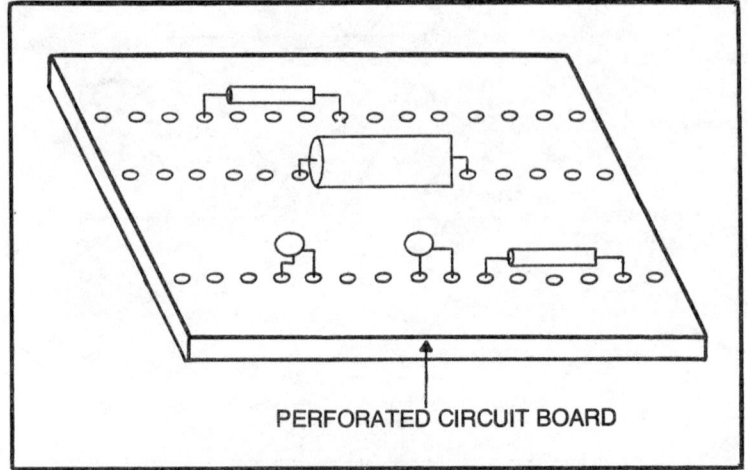

Fig. 3-7. An example of horizontal mounting of chassis components.

The second standardized method of component placement on electronic circuit board is *vertical mounting*. Shown in Fig. 3-8, each component is mounted vertically in relationship to the surface of the circuit board. Both leads are still pushed through holes in the board, but the holes used are spaced only a fraction of the distance apart when compared to the horizontal mounting method. Because the vertical space above the circuit board is utilized, the circuit board itself can be much smaller for the same circuit when compared to the board size required for horizontal mounting.

A drawback to the vertical method lies in the fact that more of the component lead is exposed on the top of the board. This could come in contact with other components and short out the entire circuit. For this reason, the lead which exits the top of the component and descends along its side to the circuit board is often covered with insulation. Hollow rubber tubing is available from most electronic hobby stores which can be slipped over the component lead before soldering it into place. When this tubing is cut to the proper length, it completely covers the exposed lead from its points of exit from the component case to entrance of the circuit board. Figure 3-9 shows how this insulating method might be used with a carbon resistor.

The more experienced builder may choose to use the vertical mounting method for constructing circuits which must be kept to minimum size. It should be remembered that only the circuit board size has been decreased. The additional space needed to house the circuit components has simply been transferred to the vertical plane. Sometimes, it will be necessary to use both methods on a single piece of circuit board, as the circumstances dictate. Should a calculation error be made and the circuit board chosen be found not to have adequate surface area for horizontal mounting after building has already begun, it is relatively easy to switch to

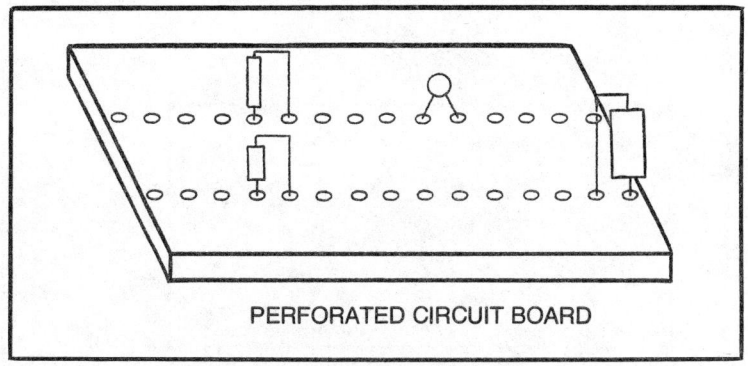

Fig. 3-8. Vertically mounted components on circuit board.

the vertical mounting method for some of the larger components to make everything fit. As long as good building practices are used throughout, this latter hybrid arrangement should bring satisfactory results.

Figure 3-10 shows the bottom of a perforated circuit board. Notice that the component leads are pushed through their various locations and then wrapped together. Components which connect to each other are placed in close proximity on the top of the circuit board to facilitate these connections without having to resort to short lengths of hookup wire. It will be necessary to use hookup wire at times, but this tends to complicate the physical circuit and is avoided where possible. Remember, the simpler your circuit is, the more reliable it will be.

After all of the components have been interconnected by wrapping them, compare what you have done with what is specified by the schematic drawing. Whether you are an experienced or new electronics project builder, you may discover that a slight error has been made. This is quite normal for all types of builders, but the ones with experience usually know

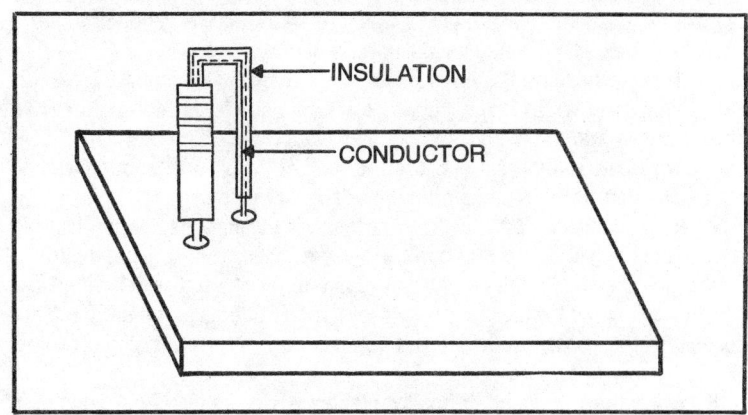

Fig. 3-9. Insulating of component lead when vertically mounted on circuit board.

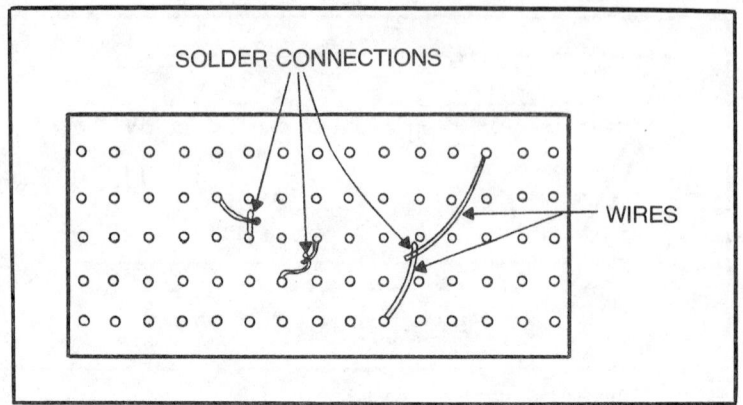

Fig. 3-10. Bottom of circuit wired onto section of perforated board.

to check and double-check all connections before soldering. After soldering is completed, all excess component leads are clipped away.

BUILDING TOOLS

For building the projects in this book, only a normal assortment of shop tools will usually be needed. One of the most handy items, that is a must for this type of construction, is a good set of small needlenose pliers. This tool is used for properly bending the wire leads of components and for wrapping them at their contact points. This wrapping is all-important, because it forms a strong mechanical connection. It will be learned later on in this chapter that a strong mechanical bond is absolutely essential before the soldering process begins. Figure 3-11 shows the lead of a component being bent at a 90-degree angle for horizontal mounting on a perforated circuit board. It can be seen that this process would be most difficult with anything but the needlenose pliers. Additionally, this tool will be used to form a heat sink when soldering delicate solid-state components which may be damaged by the heat of the soldering iron.

Another essential tool is a diagonal wire cutter, shown in Fig. 3-12. Sometimes known as diags, this tool will be used to cut away excessive lead lengths and to remove components from the circuit board. Additionally, wire cutters will also be used to remove insulation from wires and even for clipping away unused portions of the circuit board.

It is not essential to have the finest, most expensive tools available to construct the projects in this book. However, cheap tools are not desirable either. Somewhere in between, there is a midway point which offers the services required at an affordable price. The two tools just mentioned are probably the most important and most used by the electronics project builder, and therefore should be of the highest quality the reader can afford. Cheap tools may suffice for a short period of time, but sooner or later they will become defective. This is no big problem if they are

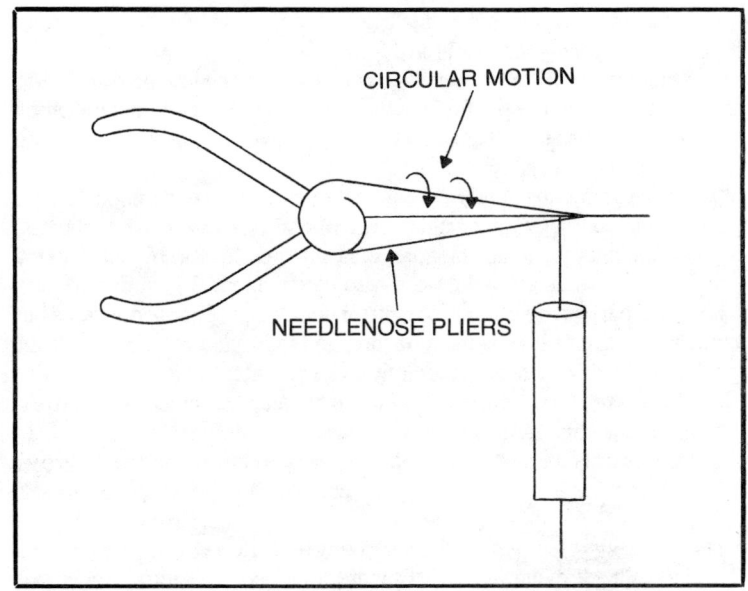

Fig. 3-11. Making 90-degree bend in component lead with needlenose pliers.

replaced immediately, but all too often the inexperienced builder will try to make a damaged tool do for awhile. This tends to create many problems, including damaged components, improperly wrapped and soldered wires, etc. Do yourself a favor from the start and purchase a reasonably good set of tools.

Another item which will be handy when building many of the projects in this book is a small pocket knife. This implement can be used for scraping insulation from the surface of painted or enamel-coated wires. It

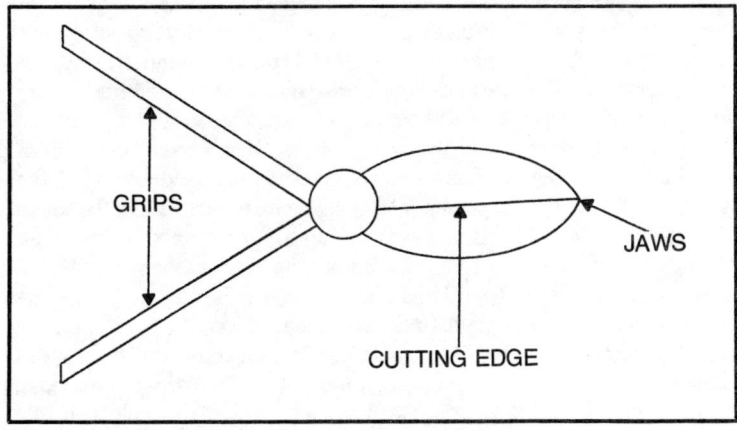

Fig. 3-12. Diagonal wire cutters.

can also be used to clear away small globs of solder which can accidentally drop between component leads on the circuit board.

Additional tools might include a wire stripper and four screwdrivers. Two of these should be the Phillips type, one of medium size and the other of the miniature variety. The other two screwdrivers should be of the same sizes but of the flathead type.

Although not absolutely necessary, an inexpensive set of nutdrivers can be useful for installing circuit board in metal cases and plastic cabinets. A nuts and bolts kit featuring a large assortment is sold by most hobby stores. This makes a good purchase to complement the nutdrivers and other tools and can be had for a couple of dollars. This hardware comes in handy when circuit boards must be mounted to insulators or where small transformers may be installed directly to the circuit board.

Accessory items which are useful to have on hand include electrical tape, alligator clips, clip-on heat sinks, and heat-shrinkable tubing. This last material can be fitted around the bare leads of components which are in danger of shorting to ground or to other components and when heated will shrink and mold itself to the conductor. Heat-shrinkable tubing is especially desirable when vertical mounting techniques are employed.

Specialized mounting may dictate the use of special tools not discussed here. It has been found that most electronics builders and experimenters have rather varied assortments of tools which can be brought into play from time to time. However, if you are buying most of the tools new to begin building the projects in this book, those items mentioned will be all that is required for average construction.

If you do not have a work bench, a kitchen table may be as conveniently used, providing that you take precautions to cover the work area so that small sections of component leads and drops of solder do not damage the finish. A vise may come in handy but is not absolutely necessary. When one is not available, a circuit board may be firmly held in the jaws of the needlenose pliers or by a pair of vise-grips. Most of the time, however, you will be turning the circuit board from top to bottom and vice versa as the component leads are fitted through one side and wrapped on the other. Figure 3-12 shows a favorite method of the author's when building circuits on perf board material. The circuit board is a bit oversized during construction to allow the edges to be placed on a couple of blocks of wood for support. Several blocks may be stacked to adjust the height of the work area. Now, the component leads may be inserted through the top of the board and wrapped without ever having to turn the board. For more stability, thumb tacks may be used to anchor the board to the wood blocks. Later, when the leads have been wrapped and it is time to solder the various connections, the thumbtacks are removed and the board is turned with its bottom side up. After soldering is completed and the excess component lead lengths trimmed away, the diagonal cutters can be used to trim the board to a more appropriate size. Since most of the projects in this book will require less circuit board size than those available through standard purchase channels, clipping away the excess will probably not be all that wasteful. This would probably have to be done anyway.

TEST INSTRUMENTS

Every electronic builder prides himself upon the assortment of electronic instruments he or she has at their disposal. Those readers with a limited amount of instrumentation will be happy to learn that the only instrument required for construction of these projects is a common multimeter. This may also be known as a volt-meter or just plain ohmmeter. There are thousands of different models on the market, but all of them measure resistance in ohms, volts AC and DC, and current. A Simpson Model 270 Meter is shown in Fig. 3-13 and is a very fine

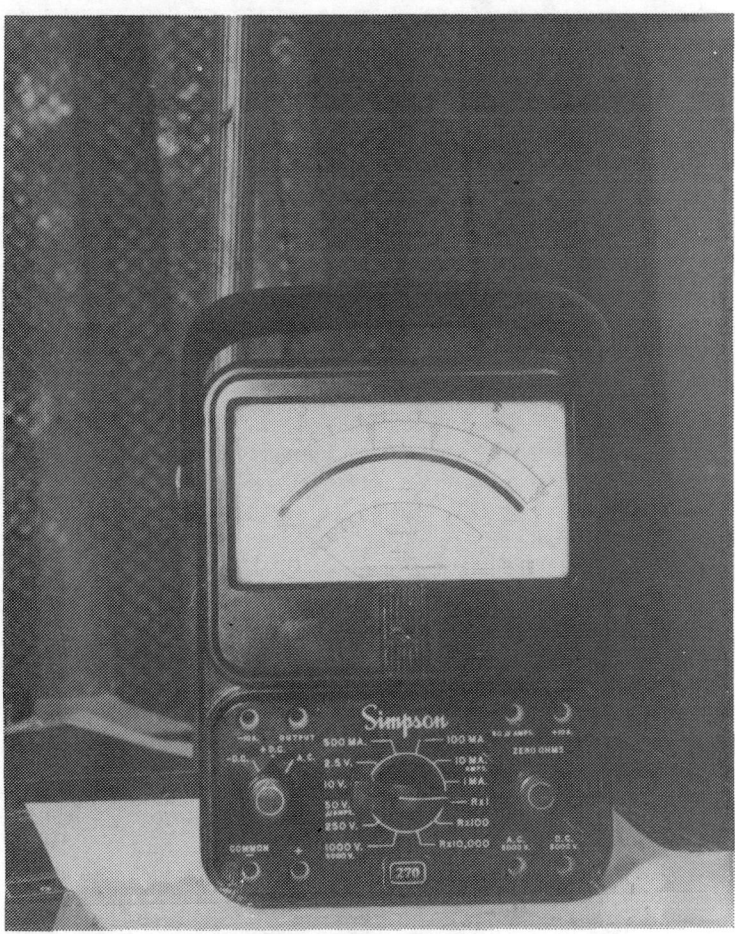

Fig. 3-13. Simpson Model 270 multimeter.

53

instrument. It will measure resistance from a fraction of an ohm to 20 million ohms. It will also measure AC and DC current, as well as AC and DC voltages up to 5,000 volts. Most other similar meters take the same measurements as the model depicted here. The big difference in price lies in the accuracy of the instrument. The Simpson Model is known for its ruggedness, dependability, a high accuracy and it is priced accordingly. Inexpensive units, which may be obtained from your local hobby store, will not offer this same degree of accuracy, but they will be adequate for the test procedures to be performed on the circuits in this book. Almost any multimeter which is operational will suffice in these applications, but don't be lulled into thinking that they will work in every instance you encounter. In applications where it is necessary to know a nearly exact measurement quantity, it must be remembered that inexpensive meters can be as far as 20% off the true reading. Again, it may be best to choose an intermediate instrument which offers reasonable accuracy at a reasonable price. Make certain that whichever instrument you choose has the capability of making fairly accurate measurements down to one-half volt. The reason for this is that the output from a single solar cell is usually around 0.45 volts. Extremely inexpensive instruments might indicate nothing at all. The Simpson instrument has a low-voltage scale which reads 2.5 volts at maximum deflection. This means that a voltage of 0.5 volts still lies fairly well up the meter's scale. As readings approach the lower end of any meter scale, they tend to become less accurate.

While some of the circuits may be further tested by oscilloscopes and other electronic instruments, in most cases the multimeter will be all that's required or even desirable. Most of the time, this meter will be used only to troubleshoot circuits which don't work on the first try. With the multimeter, batteries can be checked to see if they are good, short circuits can be detected, and open or burned out components may be identified. For most of the projects, the multimeter will not be used to anywhere near its maximum capabilities. Often, only simple continuity checks will be made with this instrument; and, of course, if the circuit works right the first time, the multimeter will not be used at all.

SOLDERING PROCEDURES

Soldering is the most important part of assembly for any electronic project, regardless of the type of circuit or the components used. One of the largest manufacturers of electronic kits in the world has stated that nearly 90% of the failures involved in putting these kits together has been traced to faulty solder contacts. A hasty job of soldering electronic components will result in a circuit which does not perform dependably or one that may not function at all. Soldering is always a critical part of electronic circuit construction and must be done with care, strictly adhering to proper technique.

The soldering iron most desirable for assembling the projects in this book is the *pencil* type, which has a power rating of 25- to 30 watts. This provides adequate heat to get the job done but does not get so hot that fragile solid-state components are destroyed. Soldering *guns* are very popular for certain types of electronic assembly, but most are rated at more than 75 watts. Soldering guns heat to temperatures which are much higher than required for the assembly of these projects. Also, the soldering tips of these guns are too large for many compact applications. Some manufacturers offer expensive soldering stations which include a pencil soldering iron, an insulated holder, and a control box which keeps the temperature of the iron constant at all times. This type of soldering equipment is shown in Fig. 3-14. While these devices are very convenient for electronic applications, they are not necessary, and a simple soldering pencil from a local hobby store may be purchased for less than $10 (Fig. 3-15). Make certain you follow the manufacturer's instructions when preparing the tip of a soldering pencil for first use. This is usually when the tinning procedure takes place and involves heating the iron and applying a small amount of solder to the tip. When the tip is covered with a very thin coat of solder, normal soldering functions may be undertaken.

Only one type of solder is suitable for use in the construction of electronic circuits in this book. This is *resin core* solder, which is sold by most electronic and hobby stores and is always identified as such. There is another type of solder which may be sold in hardware stores and plumbing outlets which has an *acid core*. The center of this solder contains an acid which is desirable for plumbing applications but which will ruin electronic circuits and components. The corrosive acid core solder usually results in cold solder joints in electronic component leads, and the acid will gradually eat away at the delicate circuit conductors. A *cold solder joint* is a connection which has not been made properly and results in high electrical

Fig. 3-14. Typical soldering station for workbench use.

resistance. High-resistance joints present most of the problems in improperly soldered electronic circuits. These bonds do not adequately conduct the flow of electrical current and can sometimes cause rectification in audio circuits. A cold solder joint is a poor or absent electrical connection. It is most often caused by simply dropping the solder onto the joint before the elements have been heated to the proper temperature. This can occur when the tip of the soldering iron is applied to the *solder* rather than to the *joint* to be soldered.

PROPER SOLDERING TECHNIQUES

Several steps are involved in forming proper solder connections when building electronic circuits. Each of these steps must be followed, in order and to the letter, to arrive at a completed product which is electrically stable and dependable. The steps are as follows:

☐ Make certain the elements to be soldered are clear of all foreign matter or debris. Wire conductors, for example, should be scraped clean of all insulation and wiped free of oil, tar, or grease.

☐ A firm mechanical joint must be formed from the elements of the joint before soldering is attempted. This is done by tightly wrapping the conductors in such a manner that no physical movement is possible between elements.

☐ The soldering iron should be of an adequate temperature to allow for proper heating. It should be turned on a few minutes before soldering is attempted.

☐ The soldering iron is applied to the joint, *not* to the solder. Once the joint has been properly formed mechanically, the soldering iron tip is placed against it to allow it to heat to the same temperature.

☐ The solder is then placed against the joint, not against the soldering iron, and allowed to flow freely around the elements. When the joint is heated to about the same temperature as the soldering iron, its elements will heat the solder and allow it to flow into every part of the wrapped conductors and contacts.

☐ Apply only enough solder to get the job done. Too much solder can create a cold solder joint.

☐ Once solder is flowing in the joint, remove the tip of the iron and make certain that the elements are not allowed to move. Motion at this point can cause the cooling solder to become cracked or loose in certain areas of the joint.

☐ Allow about 20 seconds for the solder to cool.

☐ Wiggle the protruding elements of the joint to make certain that no physical movement occurs where the solder bond has taken place.

☐ Examine the appearance of the solder joint, looking for any signs of a dull surface or globular solder deposits. A proper solder joint will have a smooth, shiny appearance; while a dull, rough surface indicates a cold solder joint.

Fig. 3-15. Inexpensive soldering pencils are available for less than $10.00.

While these steps may sound complicated upon first reading them, they will become second nature to you as you complete more and more solder connections. After only a few hours of practice, it will take only seconds to solder each joint in an electronic project. The main trick to soldering is to always apply the tip of the iron to the joint and not to the solder and to allow the solder to *flow* into the crevices of the joint before removing the heat. Remember to use the *least* amount of solder necessary to get the job done. Cold solder joints result when too much solder is used, because the cooling rate is uneven in the different layers of the molten solder which is applied to the joint elements. A soldering iron applied to the outside of a large blob of solder may cause only the outer portion to become molten while the inside remains relatively hard. This latter portion is the part of the solder joint which performs the *electrical* bonding.

A firm *mechanical* joint is stressed because solder is not of adequate mechanical strength to form this physical type of bonding. It serves only as an electrical bond, *not* as a mechanical connection. If solder is used to hold two conductors in place, for example, normal stresses will cause this connection to work loose and the solder contact to crack, since proper mechanical rigidity of the joint was not originally established. Again, solder forms an *electrical* joint. The elements of the joint must form the *mechanical* connection.

Even when using the low-wattage soldering pencils, speed in making the joint is often very important. Some of the solid-state devices used in the projects of this book can be damaged or destroyed when they become heated past their maximum points of endurance. If you are not experienced in proper soldering methods, you would do well to practice upon a more rugged device such as a resistor, capacitor, or even upon two wire conductors wrapped together. Practice proper soldering techniques until they become second nature to you. This will increase the speed with which you're able to make the joints and is important, because the longer the soldering iron is applied to the leads of a component, the hotter the component gets. A happy medium must be reached wherein adequate time is taken to complete a solder joint without taking so much time that the components become excessively heated.

A *heat sink* is often used to aid in the further protection of heat-sensitive electronic components when soldering. This is a device which *sinks* or absorbs heat. A pair of needlenose pliers can serve as a very good sink when used to tightly squeeze a lead at a point near the shell of the component, as shown in Fig. 3-16. Heat will travel up the lead from the point where it is being soldered, but the larger mass of the needlenose pliers will absorb most of it, which prevents a great deal of heat from reaching the case or shell of the component. Alligator clips and special heat sink clips can also be used to form a good source of heat protection. These devices have the advantage of remaining in place after the clip contact has been made and will free the builder's hands for other parts of the soldering procedure. When applying a heat sink, make certain that it is not located too closely to the point on the lead which is being soldered. Placed too closely, the heat sink can pull away heat from the joint and create a cold solder connection. The heat sink is best placed at a point on the lead nearest the component case. Figure 3-17 shows a heat sink attached to a transistor lead.

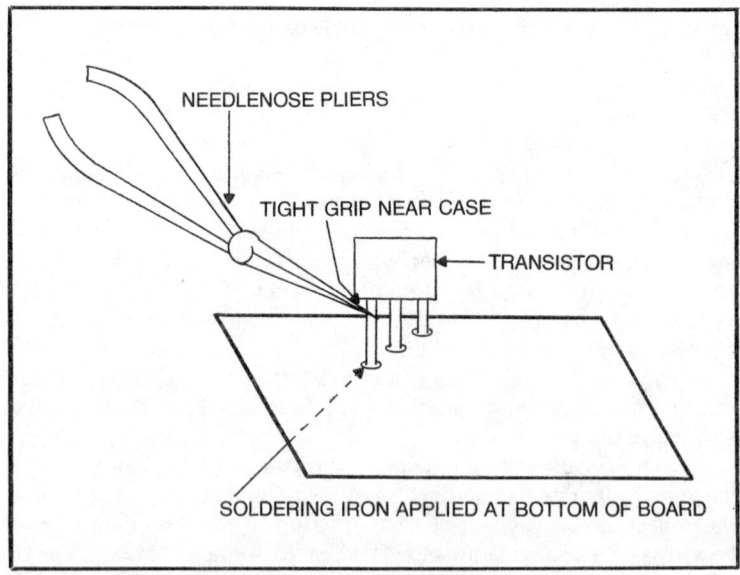

Fig. 3-16. Using needlenose pliers as heat sink when soldering.

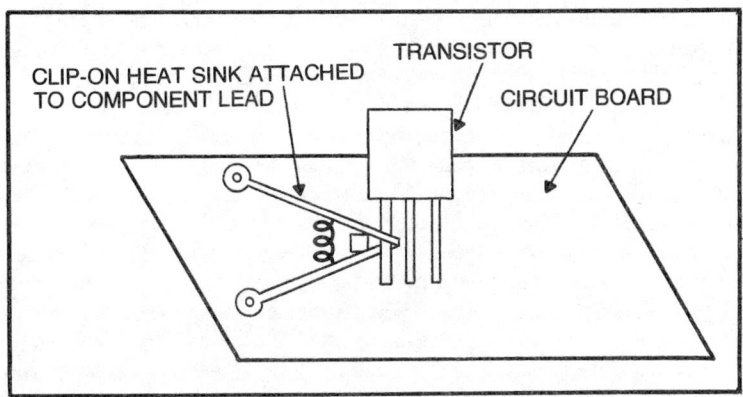

Fig. 3-17. Clip-on heat sink attached to a transistor lead.

Again, the proper soldering of electronic circuits is of paramount importance in electronic building. If you take shortcuts when putting together the circuits in this book, you are bound to run into trouble, either when the circuit is first tested, or later, when poor soldering connections break down. A few minutes spent in properly completing a project can save many future hours of troubleshooting, resoldering, and replacing heat-damaged components. Do not attempt to even start on a project until you know the correct methods of soldering.

FORMING BUILDING HABITS

At this point in our discussion, the reader should have gained most of the information needed from a technical standpoint to begin building the electronic circuits presented in this text. However, in addition to the practices and techniques described in this chapter, there are certain procedures that any builder must follow in order to be successful. These steps will help the new and inexperienced builder become proficient at the art of building electronic circuits.

Most of us have had the opportunity to see half-completed projects. Some of us may even be the one guilty of not completing what we started. In any event, these are projects which were started long ago and were to be finished up as soon as an additional part was obtained. Either the part never came or the builder lost interest, because the project was never completed. All of the time, effort, and expense went down the drain. Half-completed projects are often subject to breakage and damage because they are not in a finished form, which means they are usually not installed in enclosures which provide mechanical protection. All too often, these projects end up on the edge of the work bench and, eventually, are knocked to the floor where they are permanently damaged.

The uncompleted project is not usually the result of lack of interest, lack of ability, or lack of skill. It is often the result of beginning a project before the builder has all of the parts necessary to complete it. This is a cardinal rule of electronic building. Never begin a project until all parts, components, connectors, and the housing are on hand to complete the project. When you begin an electronic project with certain components missing, you cannot build the circuit in an orderly manner, as would be the case if all parts were on hand. The builder makes certain mental notes about parts which have been left out and which are to be replaced, and then, at a later time, completely forgets about them. A few of the components which were not on hand originally may be obtained, wired into the circuit, and then, assuming that the project is finished, power is applied. Unremembered by the neophyte builder, a component or two was omitted from the circuit, a component which the builder was supposed to have made a mental note of. Since this has been completely forgotten, the builder assumes the circuit is finished and finds that it does not work properly or at all. He now has to troubleshoot the circuit and will be more apt to look for poor connections or damaged components rather than to seek out *missing* parts. The circuit often winds up a total failure and is tossed into the junk box as a source for spare parts. Here is a good example of a circuit which probably would have worked perfectly if proper procedures had been followed.

While many builders will complete an electronic circuit having all the electrical components on hand, the case or box which is to house the finished product is often saved for last. There is nothing quite so fragile as an electronic circuit on a perforated board which is not mounted in a protective case. As soon as your circuit is completed, it should be mounted in a protective case immediately after initial testing.

A major cause of improperly wired circuits is fatigue. The author has made it a point to include this statement in every electronics book he has ever written. Experienced builders never work on circuits when they are tired, sleepy, or when their minds are on other things. Even when you are relatively fresh, it is easy to become mentally fatigued from staring too long at a concentrated area. The vision will often start to blur and hands may even begin to shake from the body being in a trance position for too long. The author calls this "builder's trance" and it is a definite trance-like state where many builders get into trouble. Most of this trouble involves the incorrect wiring of a circuit, which is bad enough; but when you are working on circuits which must be tested and adjusted while under power, this situation can be downright dangerous when medium to high voltage levels are involved.

When you feel the least bit fatigued, stop what you're doing and take a ten or fifteen minute break until you are refreshed again. Don't set a specific day or time to have your circuit completed. When running behind schedule, you may start to rush or work past your point of adequate concentration. The only result from this will be a circuit which may have problems due to polarity reversals of components, wiring errors, or improperly formed solder joints.

Make certain that the work bench area where you assemble your circuits has adequate lighting and ventilation. Arrange your seating so the normal assembly of circuits will not put you in an uncomfortable position, causing you to strain or reach in such a manner that you tire rapidly.

Most of these suggestions for good building techniques are merely common sense ideas and should be obvious to most individuals. It is a good practice to have one specific area where electronic assembly is normally done. This gets the builder accustomed to working under set conditions and makes for a more comfortable and relaxed assembly.

By following these construction suggestions, you should be satisfied each time a project is completed, both with the quality of your electronic circuit and with its operation and dependability. You should also take pride in the fact that a great many electronic projects which are built by other individuals not adhering to these techniques are going unfinished or are causing problems when completed.

SPECIAL TECHNIQUES FOR SOLAR CELLS

Solar cells are very fragile devices, as they are built around an ultrathin layer of glass. Rough handling can quickly break them, although if the surface has not been torn, they can be mended with scotch tape. Solar cells cannot be flexed even slightly. Mounting them will require special precautions in order to avoid breakage.

Solar cells should only be mounted to nonconducting surfaces which are mechanically rigid. Some of the thicker perforated circuit board materials will be adequate for the smaller cells, but the large cells will require more firmness. These latter devices will cover a large area of the circuit board and a larger area of the same material is not as mechanically rigid as would be a smaller area. One trick which can be used is combining two or more circuit boards, one on top of the other. This increases the rigidity, and the boards may be permanently glued together. Alternately, a single piece of circuit board material may be bonded to an aluminum chassis or some other firm surface to increase the rigidity of the mounting base.

Most solar cells come without wire leads. It is necessary for the builder to connect the necessary leads which are made from small sections of insulated hookup wire. One lead is soldered to the contact strip on the face of the cell, while the other connects to the metallic back surface. Figure 3-18 shows the proper contact point on a circular cell face. Any point on the back will do for the other lead.

The material used for leads must be flexible, stranded wiring. If solid wire is used, the strain from bending them during connection may cause the cell to break. Stranded hookup wire of No. 16 gauge or smaller should suffice for even the largest solar cell.

SHOCK-MOUNTING OF SOLAR CELLS

While mounting solar cells on a firm insulated surface is most ideal, this arrangement is often difficult to arrive at without special tools and

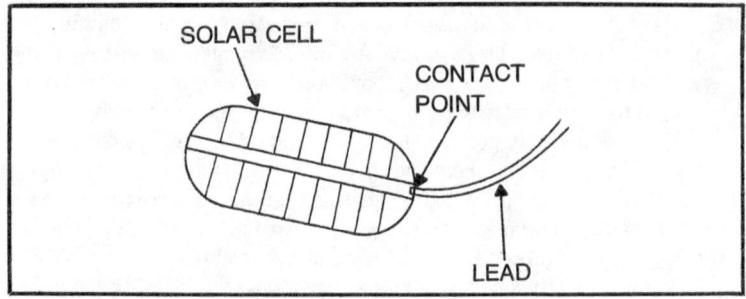

Fig. 3-18. Proper point for lead attachment on solar cell face.

materials. Large solar cell arrays will require many square inches of perforated circuit board, and if this material is attached to an aluminum chassis or other type of metal support, the weight factor can become significant. Then, too, there's the problem of the lead which is attached to the back surface of the cell. This must be brought out for contact to another cell when many are used. The small amount of solder which attaches the lead to the back plate causes the surface to be uneven, and the cell cannot be mounted in a perfectly flat attitude. In fact, if you press down on the cell too hard, the unevenness will cause the cell to break. Firm contact mounting of solar cells most often dictates the need of printed circuit board building techniques in order for the cell surfaces to be mounted in a perfectly flat configuration.

Several years ago, the author was forced to build a solar power supply from thirteen large solar cells, in a short period of time. This was done for a project which began early one morning and had to be finished before late afternoon when the sun would be nearly gone. Wiring the leads to the cells took several hours due to their size and to the extreme fragility of the cells being used. The real problem, however, came when a decision had to be made on how to mount these thirteen cells in a fairly small area without having any circuit board materials on hand. This is where what the author calls his shock-mounting system was first used. It was discovered purely by accident but works very well.

Each of the 13 solar cells used for the project came in its own special package. These were made to contain the cell while hanging on a display rack and to protect them from the normal jars and bumps that come through shipping. Each of the cells rested on a circular foam pad. These pads were brought into play for the permanent mounting of each solar cell.

Figure 3-19 shows how the actual mounting took place. The foam pad was trimmed to exactly match the diameter of the cell. The lead attached to the back of the cell was pushed through the foam pad and allowed to protrude from the edge. A drop of epoxy or one of the quick-bonding materials sold at discount stores was used to bond the cell to the foam. Another drop was used to bond the foam to a mounting surface which need not be particularly stiff. This project used a stiff piece of cardboard.

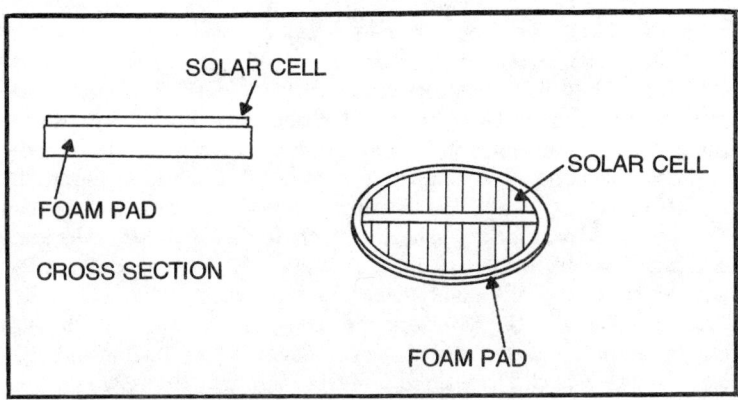

Fig. 3-19. Mounting of solar cell on foam pad.

When all of the cells were mounted in this manner, the wiring connections between them were made by twisting and soldering. The entire 13-cell circuit was mounted in a small wooden box and the project worked perfectly. As a matter of fact, this same device is still being used for various experimental purposes and no cell has ever been damaged.

This type of mounting might be considered by the reader when building solar cell circuits which are of adequate sizes to present some mechanical mounting difficulties. A test was performed with a single cell mounted on a foam pad and then the assembly mounted in the bottom of a tin can. The can was tossed about, thrown against walls, and was even slightly bent in the process; but the delicate solar cell remained intact. If the cell had been directly mounted to the bottom of the can, it certainly would have shattered from the shock of impact. For this reason, the foam mounting technique may be found especially useful when the solar cell circuit is not to be permanently fixed in one location.

Many of the circuits will not require this degree of attention to mounting considerations. A few, however, will lend themselves to this type of mounting or to similar methods. Foam pads can be cut from larger sheets of the material which may be purchased from lumber companies, discount stores, and from some furniture outlets. The material is not expensive, is easily cut by common household shears, and will conform to the size and mounting configuration of every solar cell.

Photoconductive cells, or photoresistors as they are sometimes known, do not require this amount of attention when it comes to mounting them. The same is true of phototransistors and other photoelectric solid-state devices which are packaged in a more conventional manner.

MOUNTING OF SOLID-STATE COMPONENTS

While the mounting of components in a vertical or horizontal configuration has already been discussed, it is important to go into a bit more detail regarding some of the solid-state components you will be working with during the construction portion of this book.

Figure 3-20 shows the conventional manner in which transistors are mounted through the perf board. The three leads are simply inserted through appropriate holes and wrapped with other component leads at the bottom. It is a good idea to mount transistors about a half-inch from the surface of the perf board. This will allow you to clip a heat sink to the lead being soldered near the case whenever there is a possibility that soldering will take a longer than usual time. Extreme heat for longer than a few seconds can damage these and other solid-state components.

Diodes and rectifiers may be mounted just like resistors, although due to the small size of these solid-state devices, horizontal mounting is almost always used. Diodes, too, can be damaged by excessive heating, and heat sinks attached to the two leads during the soldering process are recommended. Silicon controlled rectifiers, especially those designed to handle large amounts of current, are far more rugged and often contain large contact areas to which large conductors are often soldered in order to handle the higher current. Heat sinks are not necessary here.

INTEGRATED CIRCUIT BUILDING TECHNIQUES

Some of the construction projects in this book will use integrated circuits (ICs) in addition to the photoelectric and other solid-state components. All of the heating effects which create problems in building with integrated circuits can be overcome by using a socket. The socket is soldered into the circuit before the device is inserted and there is no possibility of any damage occurring due to heat. Adequate time may be taken when soldering these sockets without fear of heat damage. Proper soldering techniques are still dictated, however, as a cold solder joint at a socket will cause just as severe a problem within the circuit and possibly more because of the added resistance created by the friction contact of the device within the socket.

If solid-state devices are to be used with sockets, it is extremely important to make certain the device leads are cleaned of any foreign materials, especially those of an oily nature. A dirty lead can form a high resistance contact within the socket and cause the same types of circuit problems which are most often brought about by cold solder joints. The socket contacts should also be cleaned to make certain that no grit or foreign material has covered the areas which make contact with the leads. Periodic inspection of the socket is necessary, especially if the electronic circuit is used outdoors or in an area which is subject to dust and dirt buildup. A circuit which uses sockets is not quite as dependable as one which uses direct solder contact, so if high vibration applications are anticipated, the socket technique may not be practical.

Integrated circuits used in these projects are normally of two varieties, the *circular can* and the *DIP*, which is an abbreviation for Dual

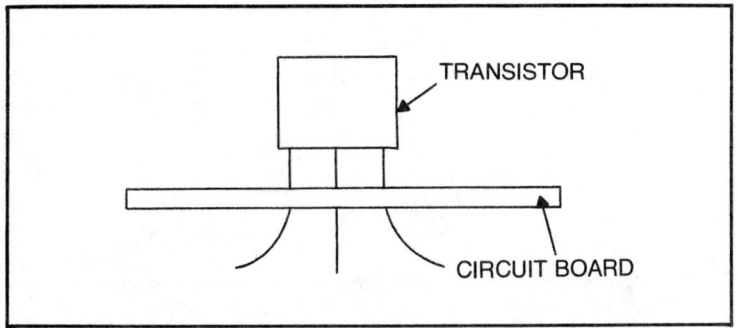

Fig. 3-20. Conventional manner of transistor mounting on perforated circuit board.

In-Line Package. There is a third integrated circuit configuration which is called a flat pack. This last type is most often used for computer applications and is very difficult to work in a typical home shop due to the extremely close spacing of the circuit leads.

Circular can integrated circuits are the type which look very much like transistors with many leads instead of just three. Often, a small tab will protrude horizontally from the bottom edge of the case to give some means of reference when determining the pin connections of the IC leads. The mounting of this type of integrated circuit to a circuit board is identical to the mounting of transistors, except more device leads must be contended with. This packaging is most conducive to the home builder because it allows for point-to-point wiring and does not necessarily relegate the builder to using printed circuit boards. When using the DIP integrated circuit, circuit boards are always required unless a socket is used which terminates in long wire leads instead of the normal pin connections. Wire lead extensions can be soldered directly to a DIP IC, but the chances are great that this process will damage the component because of heating effects. It is almost impossible to connect any sort of heat sink to the extremely short pins. Also, this packaging is usually accomplished with a plastic case which will melt and become disfigured under conditions of extreme heat. The use of a DIP socket will alleviate all of these problems.

The mounting of integrated circuits and other solid-state components when building a project in this book is best accomplished by using a small piece of perforated circuit board, which is available at most radio and hobby supply stores. It is also recommended that sockets be used for any integrated circuits which are available only in DIP configurations. Figure 3-21 shows how an integrated circuit of the circular can variety can be easily mounted on this type of circuit board by inserting each of the leads through a separate hole and then soldering from beneath. This also creates an attractive finished circuit while being viewed from the top side of the board. Many of the circular can packages for ICs contain a small plastic

65

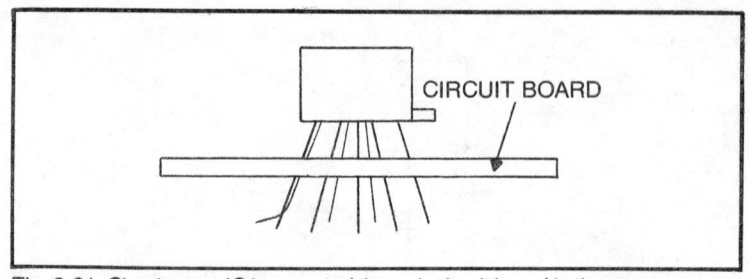

Fig. 3-21. Circular can IC is mounted through circuit board in the same manner as a transistor.

tube at their centers which acts as a divider and keeps the package slightly above the circuit board. This allows for adequate ventilation on all sides of the device housing. The perforated circuit board method of construction is technically point-to-point wiring, as opposed to circuit board construction, but the perforated board acts as an excellent base or mounting platform for all components.

Figure 3-22 shows how a completed circuit might look. Notice that a vertical mounting technique for the resistors has been used to conserve space. This is accomplished by bending the top lead of the resistor down along the side of the carbon body and clipping both leads so that the ends are even. The same is true of the mounting of small electrolytic capacitors which contain axial leads. These same components could just as easily have been mounted in a horizontal position (flush with the circuit board), if so desired. The vertical construction is intended to conserve horizontal, or circuit board, space. All connections are made from beneath by twisting various leads together and soldering them in the correct fashion.

Integrated circuits of the DIP variety will often fit in a perforated circuit board with closely spaced holes. Point-to-point wiring may be used with this type of IC if it will fit the circuit board properly, but an IC socket would be preferred. It is important to use care when installing a DIP IC in a socket. The pins of the integrated circuit are very delicate and are easily bent or even broken when forced improperly. Correct insertion proce-

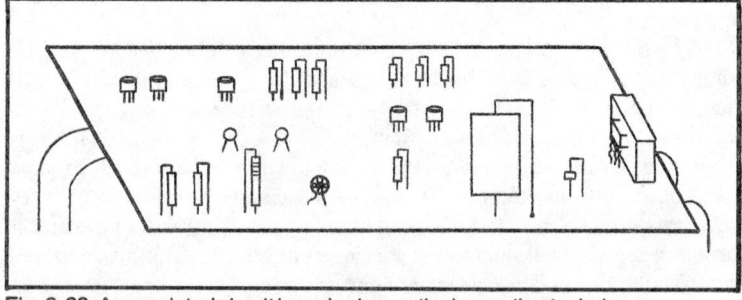

Fig. 3-22. A completed circuit board using vertical mounting technique.

dures call for aligning all the pins on one side of the IC with the holes along one side of the socket. Notice that each pin is tapered in a manner which suddenly becomes square at the midway point. Now, start each pin into its own socket hole, but do not seat them all the way. In other words, only the tip of each pin is started into its respective hole in the socket. Next, line up the pins on the other side of the IC into their respective socket holes. Make certain that the pins on the first side have not slipped from their holes while this is being done. It may be necessary to slightly bend some of the pins in order to get them to align properly. This can be accomplished with a toothpick or other small pointed device to gently force the tip of the pin into the correct slot. At this point, check all of the IC pins to make certain they are correctly inserted into each of the socket slots. Now, press firmly at the center of the IC in order to cause the remaining portion of each of the pins to snap firmly into place. A slight rocking motion when pressing the IC may cause easier entry. Removal of the device from the socket is much less complicated and is done by simply inserting a small screwdriver under one end and gently pulling upward until the IC snaps out. Practice this procedure with a defective integrated circuit if possible, because if the pins are badly disfigured, the component may be ruined.

While it may sound like integrated circuits are more fragile than eggs, this is certainly not true. They are very rugged devices; and once installed in the circuit, they provide a great deal of stability and immunity from damage due to impact shocks. The real possibility of damage is present mainly during the construction and soldering processes. Most of the dangers apply to physical breakage or destruction of the internal components due to high heating effects. Some integrated circuits are packaged in plastic cases which will quickly melt if a soldering gun is applied to a contact for too long a period of time. The real danger in heat destruction of ICs and other solid-state devices comes from applying power to a component before it has sufficiently cooled. This will often occur when defective components are being replaced with new ones. This is a simple procedure involving the removal of one device by de-soldering its leads from the circuit and soldering back into place those of the replacement component. During this process, the internal crystal chip may be heated to much higher than normal operating temperatures but not to limits which exceed its temperature seal ratings. However, if power should be applied immediately, the device might operate outside of its normal parameters and quickly be destroyed.

It is always a good idea to allow an electronic circuit a few minutes of cooling-down time after new components have been installed by soldering. It only takes a minute or so for the temperature level to stabilize, as the cases of most solid-state devices are highly efficient at getting rid of heat. Commercial solutions are available in spray cans which may be applied to heated components to quickly cool them down. These solutions are certainly not necessary for the patient experimenter who is willing to wait a few minutes before activating a newly constructed or repaired project.

Be extremely careful to bend wire leads correctly the first time, rather than having to rebend the leads to their former position and then try again. Excessive bending can easily break component leads and if this occurs at the case seal (the area where the lead enters the component case) the device may be rendered useless. If you are unsure of the correct area in which to make the bend, wait until enough of the circuit is completed to determine the correct spot, or bend the lead at a large angle rather than an acute one. The "slower" bend will not put as much strain on the lead and may be more readily corrected if necessary.

Heating of the solid-state components when soldering is a critical area, so this procedure should be done only once. The majority of components are damaged when they are incorrectly soldered into a circuit and then must be removed again because the component leads were not correctly connected at first. This assumes that the mistake was discovered before the circuit was activated, as applying the wrong voltages to solid-state leads can quickly destroy these devices. In the former case, the builder may discover when making final circuit checks that he or she has soldered an IC component or transistor into a circuit with the leads incorrectly connected. For example, some confusion over proper identification of the three transistor leads might result in the base lead connected to a point in the circuit where the collector lead should have been attached.

When incorrect solder connections have been made, it becomes necessary to heat up the leads again to remove them from the circuit. The leads are then correctly attached to the circuit points and soldered back into place correctly. If this process had been handled correctly at first, one-third of the total heating time would have been required when compared to removing the component and replacing it again. The problem would be increased even more if the incorrect connections were discovered seconds after they were made and the leads re-heated for removal and then immediately re-soldered correctly in the circuit. In this case, the device had no time to cool. After the first soldering, it was immediately de-soldered. After this was completed, re-soldering occured. The result to the component is a heating process which is three times longer than usual and nearly continuous.

The secret to proper correction of soldering mistakes is to allow the device to cool for a few minutes between each process. Upon connecting a device to the circuit, if you discover the leads have been improperly soldered to contact points, leave the connection as is and wait a few minutes for the component to cool. Then, reheat the lead and remove the device completely from the circuit. Again, allow the component to cool completely and then solder it back into the circuit in the correct manner. Allow it to cool to room temperature before power is applied.

OBTAINING COMPONENTS

Many would-be electronic project builders shy away from this aspect of the hobby, because they feel they have no sources for obtaining many of the electronic parts needed. There is absolutely no reason for anyone interested in building electronic projects to feel this way, especially when referring to the construction projects in this book. True, some projects you may see from time to time may call for an unusual part or two. Often, however, a source of supply is stated in the project description. In this book, none of the components used for completing the circuits are highly unusual; but some which are not as common as others will reference suppliers who carry them. Many parts can be directly replaced by equivalents from other manufacturers.

The place to start in being able to locate electronic components and accessories is at your local hobby stores. Here, you will find 90 percent of the items used to build the circuits in this book. These same stores will most likely carry mail order catalogues from such firms as Allied Electronics, Fair Radio Sales, and others that have been suppliers of electronic components for many years. Fair Radio Sales sells mostly government surplus materials at very reasonable prices, while Allied Electronics is a major supplier of new components and devices which range from transistors, diodes, and photocells to electronic computers and industrial electronic equipment. These are just two mail order suppliers. There are many, many others.

A trip to a local electronics repair facility can quickly get you the names of parts suppliers in your area. You may even be able to order any electronic part imaginable through these repair facilities. Once you have the name of a parts supplier, a phone call should get you a complete assortment of catalogues from the manufacturers whose products they handle. Typically, you will find catalogues from electronic component manufacturers such as RCA, Motorola, Sprague, Sylvania, Miller, Amphenol, General Electric, and many others. Just the ones mentioned here will probably be adequate to supply 99 percent of your needs. If this supplier is located in your town or city, this will probably mean same-day service if the parts you require are in stock. If you live in a small or rural area, chances are that these suppliers have route men who pass through on a regular basis. A phone call to the supplier will usually mean that your order can be delivered within a couple of days. If you elect to go the mail order route, a phone call to their order desk (often a toll-free 800 number) will allow your order to be mailed out the same- or following day.

The cost factor may drive a lot of experimenters away from electronic project building, and there is a great fluctuation in prices of equivalent components from various suppliers and dealers. For this reason, it is absolutely essential to shop around and find the best deal. The author has often replaced solid-state components called for in a circuit with a type from another manufacturer. The two parts were electronically equivalent; however, one was less than half the price of the other. This is where a lot of unnecessary expense may enter the electronic building hobby. This especially applies to solid-state devices such as diodes, integrated circuits, and transistors.

CROSS-REFERENCING

One of the most valuable aids which can be obtained from the various suppliers is cross-reference material for the manufacturers who make solid-state devices. Each manufacturer prints its own catalog or replacement guide which lists almost every transistor, diode, and integrated circuit made today and tells you which devices from this particular manufacturer are direct replacements for them. It will be necessary, or at least highly desirable, to have replacement guides from as many manufacturers as possible. For this discussion, three solid-state component manufacturers, Sylvania, General Electric, and RCA have been chosen to demonstrate device cross-referencing. Let us assume that an electronic circuit calls for a 2N403 transistor. The 2N designation is one that is rarely used anymore. It can be considered as the old generic name for transistor devices. Today, most manufacturers have an individual and specialized numbering system for naming their devices. In search of a 2N403 transistor, we first look through the Sylvania replacement guide. It reads just like a progression list. First we locate the prefix 2N and work from there. Reading up the list we have 2N399, 2N400, 2N401, 2N402 and, finally, 2N403. To the right of the 2N403 reference is the Sylvania replacement. We find that an ECG 102 is a direct replacement for the 2N403. The same can be done using an RCA catalog, which will tell us that the SK 3003 is a direct replacement for the 2N403. It is also a direct replacement for the ECG 102. The General Electric replacement guide specifies a GE-53 as the exact replacement for any of the devices mentioned so far. If the transistor number supplied in the circuit diagram were that of another manufacturer, this device could also be referenced to the current device equivalent from another manufacturer. Assume that a transistor is called for which is a Sylvania ECG 107. The RCA replacement guide would list the ECG 107 along with the RCA equivalent, which is an SK3293; or the General Electric guide would specify a GE-11. All of these devices are electronically equivalent. The great majority of them will contain the same case and lead configurations; however, if is often good to make certain that the leads are arranged in identical order. Sometimes, different cases will be used to house an equivalent crystalline chip; but these occurrences are usually specified in the cross reference manuals.

The vast numbers of transistor identifications often confuse and frustrate both beginning and experienced electronic experimenters. If you keep the cross reference manuals on hand at all times, you may find that you have an equivalent device already on hand, either new or contained in a junk box circuit.

THE EXPERIMENTER'S JUNK BOX

Every able experimenter accumulates a "junk box" of parts in a very short period of time. The term junk box is a misnomer, as many of the components contained there will be put to good use in future projects. A junk box may actually be several crates, boxes, and cabinets filled with surplus transmitters, receivers, old data circuit boards, electric clocks, etc. Through many surplus outlets, it's possible to obtain brand-new circuit boards from discontinued lines. Manufacturers dump a lot of outdated components and circuits into these industrial surplus channels and some very good buys can be made. It is not uncommon to purchase a transistor or IC from one of these mail order dealers which is guaranteed to be new and operational and which sells for one tenth of the price of an equivalent component on the retail market. Rectifier diodes may be purchased for 11 cents apiece; the same device might cost $2.50 if purchased from a retail repair facility. Recently, when several citizens band radio manufacturers went bankrupt, huge numbers of their completely wired circuit boards were available through surplus channels. They sold for less than $10.00 and contained close to $100.00 in brand-new parts. With a soldering iron and some desoldering tape, most of these components can be salvaged in like-new condition. The leads may be a little short, but the low price is worth the extra difficulty encountered in the salvage.

Enterprising experimenters keep track of these products through ads placed by the distributors and dealers in electronics magazines. They also send a letter to the distributors asking that they be placed on a regular mailing list to receive constant updates on the products that are available. When a good buy comes along, one which offers components of immediate or future use, a purchase is made. These parts are stored away until needed. This is what constitutes a junk box. Sometimes, a buy is made simply due to low price without any idea of what the purchase may be used for. Intuition plays a role here, and occasionally these purchases turn out for the best. Often, however, the components bought sit around for a few years and are traded off to other experimenters, resold, or tossed in the real junk box, often called the trash can. In the early stages of building up a parts supply, it is very easy to become enamored with having a lot of *anything* electronic, regardless of whether or not it would seem to have a future use. This is an excellent way to waste money. Don't become so parts-conscious that you must have everything that is available at a ridiculously low price. Some components are this cheap because their practical uses are ridiculously few and far between.

There is another aspect to building a junk box which costs little or nothing. Many amateur radio operators, CB enthusiasts, and experienced experimenters have junk boxes which have been transformed into junk rooms. Their amount of junk has reached a stage where it is overflowing the owner's ability to keep it in an orderly fashion. In these situations, it may be possible to haul away wheelbarrow loads of components for free. This is more preferable to the owner than having to pay to have it hauled to the junkyard. Don't neglect your local radio and television stations either, nor nearby television and stereo repair shops. One might be surprised at the wealth of electronic components which may be obtained from an old television receiver. True, it is a tedious and often dirty job to remove all of the components, but often worth the effort.

The author has been successful in salvaging many electronic parts from printed circuit boards in a rapid fashion by cutting the circuit board out from under its components. The only tools required are diagonal cutters and needlenose pliers. Most of the older types of circuit board material will chip away in large pieces when cut. The green, glass-epoxy boards will present major problems with small cutters and a larger tool might be brought into play. Usually, the entire circuit board can be cut from the components without having to use a soldering iron. Gobs of solder which remain at the ends of the component leads turn to dust when tightly squeezed in the jaws of the long-nosed pliers. Medium-sized circuit boards with 50 to 100 components can be cleared within a half hour to forty-five minutes. True, a few components do get damaged from this rather hasty method, but any especially critical devices may be removed ahead of time by de-soldering the leads in a conventional manner. Clip-on heat sinks can be used to protect them if necessary.

Your local electronics hobby store may also be a source of almost free junk box parts. The local Radio Shack store in the author's city often displays defective or damaged merchandise on a special counter which can be had for pennies on the dollar of retail cost. Additionally, discontinued items may also wind up here at reductions of over 75 percent. Many of the damaged items can be repaired, while others can be salvaged for whole circuit sections or into individual components.

Don't neglect the mechanical parts either. These will be found through some electronic outlets and in the junk boxes of other experimenters. These include switches, relays, and mechanical timers which may often be put to good use in various electronic circuits. Buying these parts new could cost $50.00 or more. This is ridiculous considering the fact that a surplus component might suffice just as well. It might even be better. One has to realize that government and industrial surplus parts may have originally cost hundreds or even thousands of dollars and may sell for less than ten dollars. The price differential quoted is not a highly unusual one in the world of surplus components and will be verified quite often as you continue to experiment and obtain parts.

KEEPING TRACK OF ELECTRONIC COMPONENTS

Once you have built up an adequate supply of electronic components, the problem emerges as to how to keep track of everything you have. The reason for stocking up in the first place was to allow you more flexibility in being able to tackle simple electronic projects without having to order every part required. But if you can't lay your hands on what you've collected or are not sure of all that you have, the original purpose for obtaining the parts may be lost.

Fortunately, it is very simple and inexpensive to categorize the great majority of your surplus components and the entire process need take no more than a leisurely weekend. To repeat, parts categorization is absolutely essential to the operation of a successful experimenter's workbench.

First of all, begin with the heavy components. Large transformers, chokes, and similar devices may all be placed in a wooden or cardboard box and stored in a closed cabinet or on a shelf. The side of the box should be plainly marked to identify the materials it contains. If you have a great deal of different transformers, perhaps all of the low-voltage units could be in one box while the medium- and high-voltage parts are placed in two others. Each would be labeled accordingly. Cigar boxes make excellent storage trays for small components. Small audio transformers, switches, relays, etc. may be kept here. The front or side of the cigar box is marked with a black pen to identify the contents. While it is a good idea to house only one type of component in each box, it is also possible and practical to keep a variety of components which are interrelated as to their uses in circuits. For example, one box might be marked "AM Radio Parts" and might include capacitors, ferrite rod antennas, speakers, etc. that were removed from AM radios. Cigar boxes can be stacked, one above the other, and contained in a small space. They are very sturdy and will last for many years if stored away from damp areas.

For the individual components such as resistors, capacitors, and solid-state devices, workbench shelves with plastic racks may be purchased. These are available from discount stores and many other sources. Shown in Fig. 3-23, a typical cabinet may contain from twelve to over forty plastic trays which slide forward to give access to their contents. Transistors may be categorized and placed in each drawer. Often, these drawers can be subdivided into three different compartments

Fig. 3-23. Workbench parts-storage cabinet.

Fig. 3-24. Plastic small-parts cabinet designed for stacking.

using plastic spacers contained with each purchase. Three different kinds of transistors, diodes, capacitors, etc. can be stored in this single drawer and kept separate from one another. Figure 3-24 shows a small parts cabinet which is especially designed for stacking purposes. This is more apparent when looking at the top and bottom of the unit which is shown in Fig. 3-25. The bottom is recessed and will slip over the raised, mating surface which is found at the top of a similar unit. Figure 3-26 shows how several units may be stacked to conserve space on a workbench.

Resistors should be stored according to ohmic value. Figure 3-27 shows a resistor storage cabinet which is commercially manufactured and contains several different compartments in each drawer. Each of these compartments has been marked with a value. Fortunately, ohmic values of carbon resistors are pretty much standardized. For example, there is a 2700-ohm and 3000-ohm resistor, but none are commonly manufactured for 2800 ohms. Because resistance values are standardized, it is quite simple to purchase a resistor cabinet which is marked to identify common carbon resistors. Of course, if you prefer, parts cabinets similar to the ones previously discussed can be marked with resistance values and used in place of a commercially manufactured cabinet.

It is also a good idea to equip your workbench with several large plastic trays which can be used to house various "pull-out" components until they can be properly categorized and stored. Always discard damaged components or those which are never to be used. These have a tendency to build up and become intermixed with the categorized components. The job of keeping an orderly parts system is mainly comprised of throwing away what you don't need, rather than keeping what you've already filed in a constant state of order. If you do the former, the latter will usually take care of itself.

If dry cell batteries are to be kept, they should be housed in metal drawers or boxes. Make sure these compartments are leak-proof. Battery storage trays should be checked periodically in order to identify and discard any cells which may be leaking. These can quickly damage the remaining good cells and a leak can ruin other components and finished circuits.

This storage system should encompass the majority of the parts and components you are likely to collect. There will always be some devices and circuit parts which cannot be accurately filed away, and these must be relegated to a "Miscellaneous" box or cabinet. Whenever possible, the

Fig. 3-25. Bottom and top views of previous cabinet showing mating surfaces.

materials filed under miscellaneous should be placed in a more appropriate category as time permits.

With this type of filing system, it should be much easier to lay your hands on the parts which are needed to build a proposed electronic project. Once you know what you have, it is an easy matter through process of elimination to determine which components you will have to purchase. Once these are on order, your project is well on the way to getting off to a good start.

SUMMARY

The successful completion of an electronics project is dependent upon a correct start, adherence to proper building techniques, and the availability of the proper parts. Putting together a project is not highly complicated and should be pleasant and interesting rather than straining one's patience and mental processes. If all is in order before you begin, all parts on hand and all tools readily available, construction should go like clockwork. On the other hand, if some parts are missing and your complement of tools insufficient, you may be taking on something that will cost you a lot of time, money, and frustration, while delivering nothing in the end.

Never push yourself. This is a cardinal rule. *Never* give yourself a deadline for completion of a project. This will tend to cause you to hurry, especially if progress is slow at first. When hurried, builders (experienced and inexperienced) make *mistakes*. Mistakes cause damaged components, inoperational circuits, and sometimes the scrapping of the entire project. Chances are, your life and future happiness do not depend upon the completion of a successful project, so why rush to finish something that is for your enjoyment and not absolutely essential. By proceeding in an

Fig. 3-26. Former cabinets in stacked configuration.

Fig. 3-27. Commercially manufactured resistor-storage cabinet.

orderly and logical fashion, you will learn more about the building processes, the components, and the circuit function than you will by hurrying. Also, your circuit stands a 100 percent better chance of operating properly on the first try.

While we are talking about simple circuits in this particular text, the reader must realize that construction habits are most often formed during the building of simple circuits. If this learning process is handled in an incorrect manner, bad habits will have to be broken at a later time. Proceed as directed and all of your attempts at completing a functional electronic circuit will certainly meet with success.

Chapter 4
Electronic Solar Projects

Having discussed the tools, components, and building practices which will be involved in building electronic solar projects, it is now time to move on to the projects themselves. This chapter features 33 projects of varying degrees of complexity, all involving solar cells, photoresistors, phototransistors, or combinations of photoelectric devices. Wherever possible, suggested manufacturer's part numbers are supplied, especially when the component may be a bit unusual or difficult to locate through normal channels. Most applications of solar cells in these projects, unless otherwise noted, are noncritical as to the type of device chosen. In most of these circuits, only low-powered solar cells will be required; however, large, high-current devices will also work as well but at a much higher cost factor. This does not apply to certain solar-powered supplies which will require solar cells of the minimum current rating specified in order to perform as projected.

In most projects, a parts layout or mounting diagram is included along with the schematic drawing. This will help you to get a rough idea of where the parts might be positioned for most convenient wiring on the circuit board. A few projects are mounted directly in plastic cases and do not require circuit board. Here, case-mounting diagrams will be offered.

If you are in doubt as to some of the schematic symbols used in the circuit drawings, you may refer to Appendix A, which provides a schematic symbols chart that will aid you in component identification. Feel free to substitute components when necessary, although staying with direct replacement parts will aid the inexperienced builder who may not yet have the knowledge to perform the slight wiring or associated component modifications which might be dictated by substitution of a device vital to a circuit which has not been provided for in the original schematic drawing.

Although it has been mentioned before, it will be repeated: Do not begin construction of any electronic project until you have *all* of the necessary parts and components for completion. Start by ordering all of the parts and then study the schematic and building plans in your mind until all components are on hand. Chances are, the parts needed for most of these projects will be available in your home town or neighboring city. No device is terribly expensive and should be within the economic range of most every builder.

TROUBLESHOOTING

Each of the circuits provided in this chapter has been checked for accuracy. When assembled as outlined, each should perform closely to its stated operational characteristics. It should be pointed out, however, that differences in components made by different manufactures can, from time to time, cause certain circuits to operate a bit differently than would be expected. This should not be the case with the majority of the circuits found in this book; however, if certain discrepancies should occur, it may be necessary to go through the various components with meters to make certain their values are correct. Some surplus components may have false values marked on their cases. This is a very rare occurrence but can lead to difficulties in electronic circuits.

The chances of obtaining a falsely marked component or one that is significantly different from that specified in the schematic due to difference of manufacture are very rare and not really worth worrying about. The majority of inoperative circuits are so because of faulty solder joints. These can be completely eliminated by adhering to the building practices discussed earlier in this book. The next highest cause of circuits which refuse to work properly or at all is improper wiring. This means that wire A, which was supposed to have been connected to contact B, was inadvertently connected to contact C instead. The circuit has been incorrectly wired, although soldering, component placement, and actual construction practices have been closely adhered to. All but one. Before making a solder connection, you should always double-check the schematic drawing to make certain that what you are about to do is correct. Once a project appears to be finished, all wiring connections should be closely examined and referenced to the schematic drawing before power is applied.

The third highest cause of project failure is due to defective circuit components. This would include bad transistors and integrated circuits, broken wires, and burned-out resistors. An ohmmeter can be used to detect many problems that may develop with all of the devices used for construction in this book.

As you continue to build electronic projects, you will eventually have to troubleshoot some of the circuits after building is completed. Mistakes will be made and tracing them down through the process of elimination can be a very educational experience.

PHOTOVOLTAIC LIGHT METER

Camera enthusiasts have long used light meters to determine the f-stop settings and shutter speeds of their cameras. Many of the early light meters consisted of a circuit similar to the one shown in Fig. 4-1. They used a small photovoltaic or solar cell in combination with a simple meter and, often, a variable resistor which was mounted internally and used to align the meter when referenced to a known light source.

The circuit in Fig. 4-1 can be used to check the level of various types of light, both natural and artificial. It is a very simple device and can be constructed in less than an hour for about five dollars if you can locate a surplus meter. By referring to the schematic, it can be seen that the circuit is simply a power source (the solar cell) connected to a microammeter with a series resistor that can be varied to limit the current flow.

When the solar cell is subjected to a specific level of light, current begins to flow in the circuit. The variable control is used to reference the pointer in the meter. This will most often be set at mid-scale, or 25 microamperes. Now, with the variable control in the same position, other levels of light can be measured in reference to the first source. With one of these new sources, a reading of 12 microamperes would indicate an intensity of brightness of about one-half that of the reference source. A full-scale reading of 50 microamperes would indicate twice the intensity of the reference source. This assumes that the reference light is of the same frequency as the other two.

The inset in the schematic drawing shows how the 50,000-ohm potentiometer is wired in the circuit. It is actually used as a variable resistor rather than a potentiometer which controls resistance in two directions or in two circuits. The variable resistor controls resistance in only one circuit leg. A potentiometer has three contacts, as is shown in the inset, but only two are used. By looking at the device with the shaft facing you, the contact on the right and the one in the center will be the ones normally selected for this type of application because as the control is rotated clockwise (from left to right), the meter indicator will tend to rise in the same direction (from left to right). This is because the resistance decreases within the potentiometer when using these two terminals and when the control is turned in a clockwise position. When the shaft is rotated in the other direction (counter clockwise), the meter needle will also move from right to left, dropping down its scale. If the left-hand and center contacts were used instead of the right and center combination, the control would have the opposite effect on the meter. A clockwise rotation would cause the meter reading to decrease in value, while counterclockwise would produce a lower reading. Either set would work well, but most of us are accustomed to obtaining more of something when we twist the control from left to right (e.g., radio volume controls or light dimming controls).

The two outside contacts of the potentiometer would never be used alone. If they were, turning the shaft would have no effect on the resistance

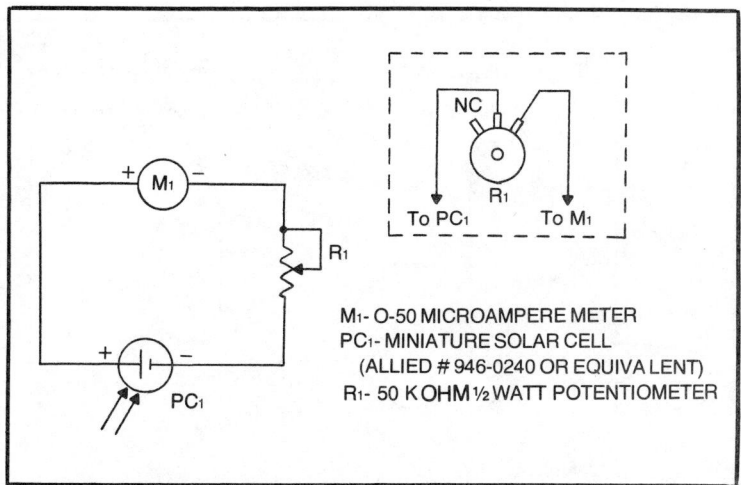

Fig. 4-1. Photovoltaic light meter circuit.

whatsoever. Using this particular control, the resistance between the two outside terminals will always be the highest value of the component, or 50,000 ohms. This value is not changed by rotating the shaft. So, a variable control will *always* be used with its center contact connected to a circuit, and at least one of the other contacts also utilized, depending on the specific application.

Figure 4-2 shows a pictorial drawing of how the circuit is mounted in a small plastic or aluminum case. Plastic is preferred because no insulator will be required for the solar cell whose wiring must travel through the case to the meter and variable control. The solar cell is mounted to the top of the rectangular case and may be held in place with a drop or two of epoxy cement. The meter is mounted through the front face and a half-inch hole is drilled for insertion of the variable control which is held in place by a nut and washer. Drill two small holes in the top of the case and run the solar cell wiring down to the internal components. The positive lead connects to the positive terminal of the meter, while the negative terminal connects to the center contact of the variable control. A piece of stranded hookup wire is soldered to the right contact of R1 and is connected to the negative meter terminal. The project is complete.

Check-Out Procedure. Check-out is very simple, as your circuit may begin to give meter indications when the last wire is connected. Start by turning the control full counterclockwise. Now, hold the case so that the illumination from a nearby light bulb is concentrated upon the solar cell. Slowly adjust R1 in a clockwise direction until a reading is obtained on the meter. Place your hand between the light source and the solar cell, casting a shadow on the cell. The meter indication should drop significantly. That's all there is to it.

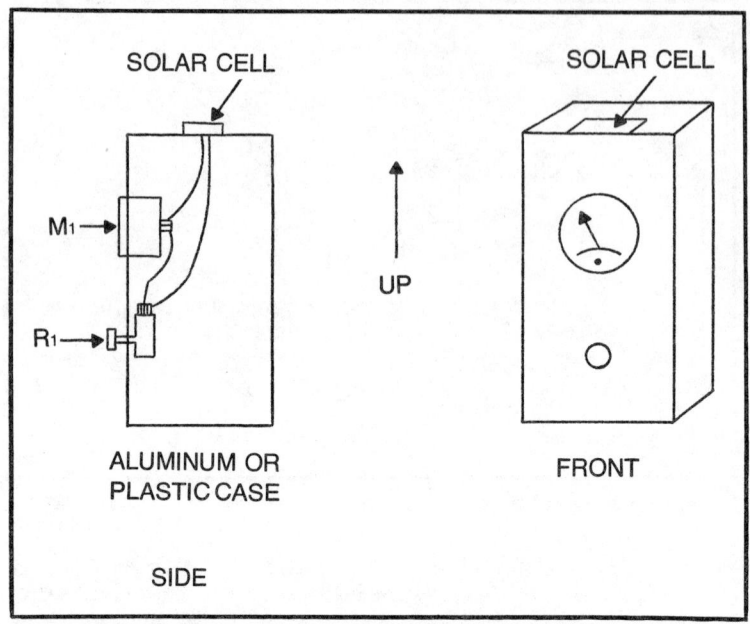

Fig. 4-2. Mounting of light meter in plastic or aluminum case.

This device can be calibrated by using an adjustable source of light whose intensity in candlepower is already known. A pointer-type knob can be placed on the variable control and a scale can be drawn around the shaft. This will allow you to set the meter to read full-scale at a certain intensity. Alternately, the variable control can be locked in one position and the meter readings observed and noted for many different levels of intensity. These readings can be posted on a small chart taped to the back of the case.

PHOTORESISTOR LIGHT METER

Another form of light meter which will accomplish the same purposes as the one just discussed is shown in Fig. 4-3. This circuit contains its own internal power supply, a 1½-volt "AA" battery. A photoresistor is used as the light source indicator and changes its resistance and thus the current flow through the DC meter, which has a 1 milliampere movement. The type of photoresistor used is not critical, although it should exhibit less than 500 ohms resistance in bright light. The Radio Shack unit used in the test circuit has a minimum resistance of about 100 ohms and a maximum of 5,000,000 ohms in the dark. This component provides excellent response and sensitivity over a wide power range.

The 1½-volt battery supplies constant current for this circuit. The current flow is referenced as before with a variable control wired in the same manner as with the circuit in Fig. 4-1. A single-pole/single-throw toggle switch is used to conserve battery life. This was not necessary for the previous circuit, as turning the sensitivity control to full counterclockwise effectively shut down the circuit and there were no components which stored power, as is the case with this circuit. When the meter is not in use, the switch is turned to the off position. This also helps to prevent accidental meter damage by subjecting the photoresistor to an extremely bright light source inadvertently while transporting this device.

Construction is handled in almost the same manner as before, but two additional components must be added to the interior of the case. Another one-half inch hole is drilled beneath the one used for mounting the variable control. If a sub-miniature switch is used, a quarter-inch hole will probably be all that's required. The manufacturer's specifications with the switch will determine the mounting hole diameter. Inside the metal or plastic case, a battery holder must be mounted which will keep the "AA" battery in place and will provide solder contacts for connection to the rest of the circuit. The type holder used for the test model of this circuit was a Radio Shack #270-401, which sells for $.39. This holder is secured in the back of the case with epoxy or some other bonding compound. Make certain none of the compound is allowed to drip across the battery, as it may be permanently bonded to the holder and replacement will be impossible without replacing the holder as well.

Figure 4-4 shows front and side views of the circuit and case. The photoresistor is mounted on top and its leads connected as indicated. If an aluminum case is used, insert two rubber grommets in the drilled holes

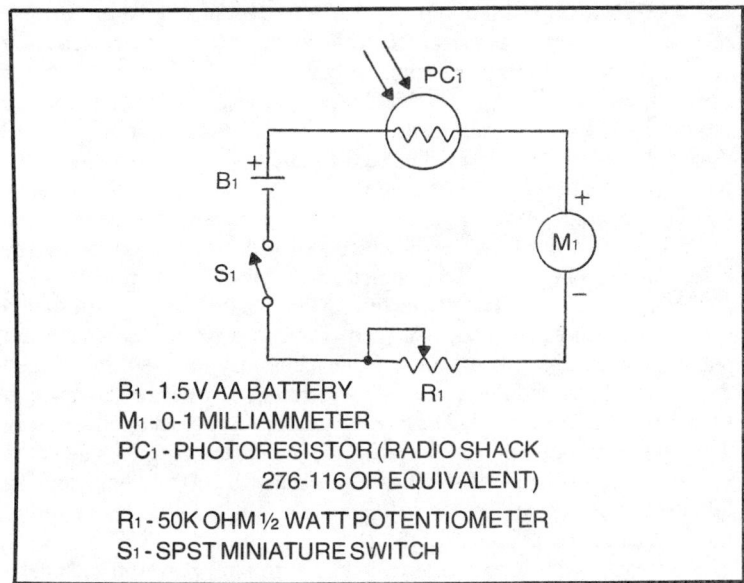

Fig. 4-3. Photoresistor light meter.

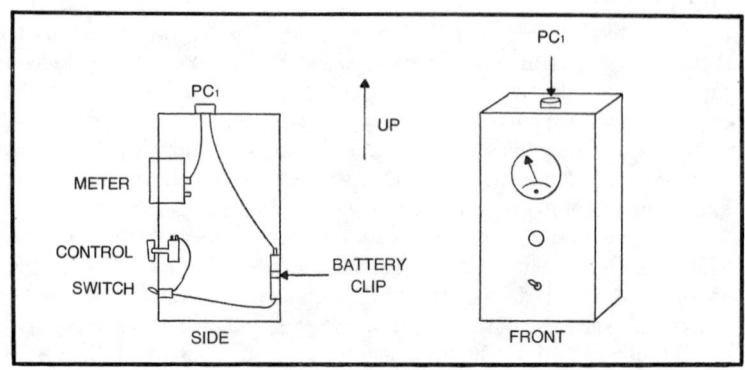

Fig. 4-4. Front and side views of photoresistor light meter case.

through which the photoresistor wires are run. Insulation on the wires can be cut by the sharp edges created when holes are drilled in aluminum. Grommets are very inexpensive and can be purchased in packs of 100 or more for less than a dollar.

Once everything is installed, look over your wiring connections carefully and turn the switch to the off position and the variable control to full counterclockwise *before* installing the battery. If this is not done, the meter movement may be damaged by the current within the circuit should the photoresistor be exposed to an extremely bright light. Insert the battery and close the case. Your project is complete.

Check-Out Procedure. With the variable control still in a full counterclockwise position, flip the switch to the on position and expose the photoresistor to a source of light. Slowly rotate the control in a clockwise direction until a reading is obtained. Place your hand between the light source and the photoresistor. The reading should deviate as you move your hand back and forth. If the meter reading tends to travel downward below the zero reading on the scale instead of up the scale, the polarity is incorrect. This can be fixed in two ways: the terminal connections at the meter can be reversed; or the battery may be reversed in its holder. Any other circuit difficulties will probably be caused by broken or improperly wired connections or by a discharged battery.

Note: All of the light meter circuits in this chapter have been designed using the specific photocells indicated. The photovoltaics have a current output of about 10 milliamperes. Other cells may be used, but if they deviate significantly in output, some adjustment may have to be made to the value of the variable resistor. If the control is very touchy and suddenly goes from zero to a full-scale reading, this indicates that the value of the variable resistor should be significantly decreased. On the other hand, if a full counterclockwise position of the control does not completely return the meter to zero, the value of the variable resistor should be increased. This probably will not be necessary unless extremely powerful solar cells are used.

COMBINATION LIGHT METER

For those experimenters who would like to go a bit further, the circuit in Fig. 4-5 offers a little more complexity but can still be assembled within an hour. It takes advantage of two types of photoelectric devices; the solar cell and the photoresistor. Because two photoelectric devices are used, this circuit is very sensitive to light variations while still maintaining its simplicity. It does not require a battery power source and is always operational.

Care must be taken to make certain the polarities of the solar cell and the meter are aligned. The photoresistor and variable control are not polarity-conscious and may be connected in any manner. Figure 4-6 shows the plastic or metal casing used to house this circuit and the mounting location for the two photoelectric devices. Four holes will have to be drilled to accommodate the photocell leads. Again, epoxy is used to permanently bond these two devices to the top of the case. The positive lead from the solar cell is connected to the positive terminal of the 50-microampere meter. The negative lead of the cell connects immediately to one of the leads from the photoresistor. These two wires should be insulated if an aluminum case is used. Solder the connection between the two cells. The remaining lead of the photoresistor is soldered to the center contact of the variable control. Turn the control to a full counterclockwise position before making the last connection by soldering a piece of hookup wire to the righthand contact of R1. The other end of this wire connects to the negative terminal of the microammeter. The circuit is now complete, and the case may be secured.

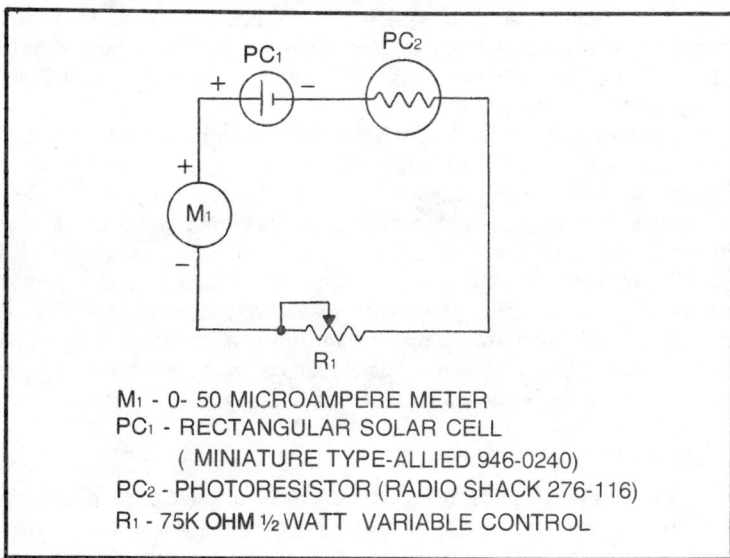

M_1 - 0- 50 MICROAMPERE METER
PC_1 - RECTANGULAR SOLAR CELL
   (MINIATURE TYPE-ALLIED 946-0240)
PC_2 - PHOTORESISTOR (RADIO SHACK 276-116)
R_1 - 75K **OHM** ½ WATT VARIABLE CONTROL

Fig. 4-5. Combination light meter circuit.

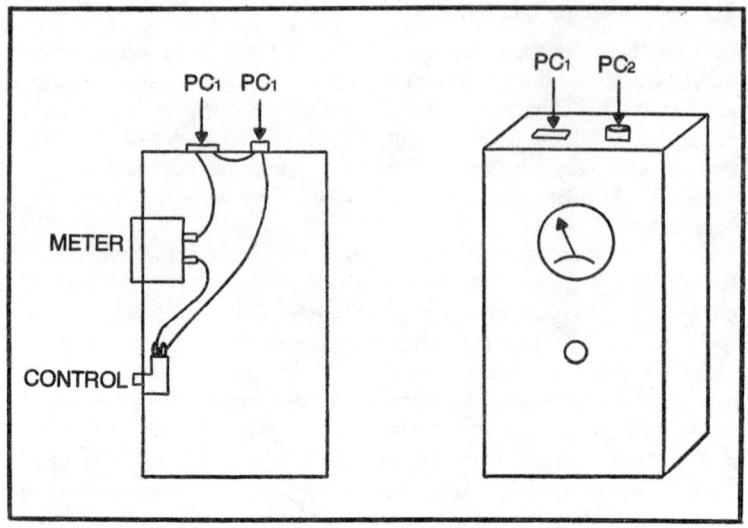

Fig. 4-6. Mounting of combination light meter circuit in plastic or aluminum case.

Check-Out Procedure. Just as before, expose the photocells to a moderately bright light source and adjust the control until a reading is obtained. If the reading varies when you cast a shadow on the cells, the circuit is working properly. When not in use, the meter should be protected by turning the variable control to the full counterclockwise position. Also, protection can be obtained by covering the photoresistor with a metal thimble or other cup-like object which will keep all light away from its treated surface. Using the device specified, circuit resistance will increase to about 5,000,000 ohms, and even the brightest light will not cause the meter to move.

The meters specified for the three light-meter projects discussed in this chapter may be altered significantly, probably without having to change the values of any other components. Common movements have been chosen, but if you have other microammeters or milliammeters on hand, they can probably be used. Once a meter exceeds 5 milliamperes as its full-scale reading, its sensitivity when used with small solar cells may decline to a point where the device is not very useful. The battery-powered circuit should work reasonably well with milliammeters which have even higher full-scale readings. Experiment with the meters you have on hand and change the value of R1 accordingly, if necessary.

THREE-VOLT POWER SUPPLY FROM THE SUN

It is possible to power many different types of low-current electronic circuits directly from the power of the sun by using low-cost miniature solar cells. Normally, these must be stacked to provide the voltage rating needed by transistors and other miniature solid-state devices. Figure 4-7 shows a circuit which uses seven rectangular solar cells in series to arrive at an output voltage of 3 volts DC. The solar cells used provide a current output of about 20 milliamperes and can be purchased for $1.20 each if you buy 10 or more. While only seven are used in this circuit, the other three could certainly be put to use in other electronic solar projects.

The cell specified for this project measures 0.8 x 0.2 inches and can be easily mounted to a small rectangular section of printed circuit board. Since each device has an output voltage of 0.45VDC, seven of them in series should produce 3.15VDC. However, a diode is placed in series with the output to prevent the cells from becoming damaged should they be used to charge other batteries. This component will create a voltage drop, lowering the output to about 3 volts. Without this diode, an overcharged battery could conceivably feed power back into the solar cells and destroy them. The term *other* batteries was just used, because this circuit is often called a solar battery. This is especially true when this circuit is connected to a rechargeable battery stack such as the nickle-cadmium type.

The solar cell power supply may also be used to supply operating power to small radios, miniature transmitters, and many other electronic circuits which will operate from a 3-volt supply delivering 15 to 20 milliamperes of current in bright sunlight. If higher current is needed, two rechargeable 1.5 volt batteries may be used as the main power source with the solar cell supply acting as a battery charger. In this way, the circuit receives its power from the batteries; but these are charged by the solar cell supply when the circuit is not in use.

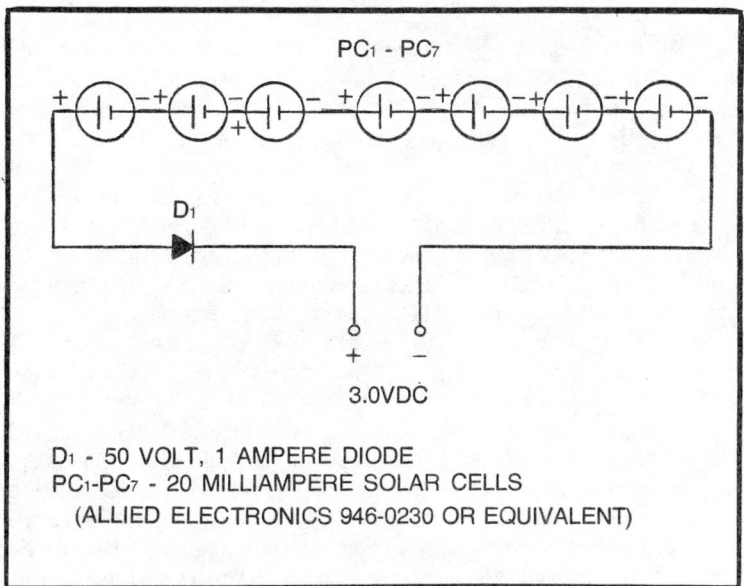

Fig. 4-7. Three-volt solar power supply circuit.

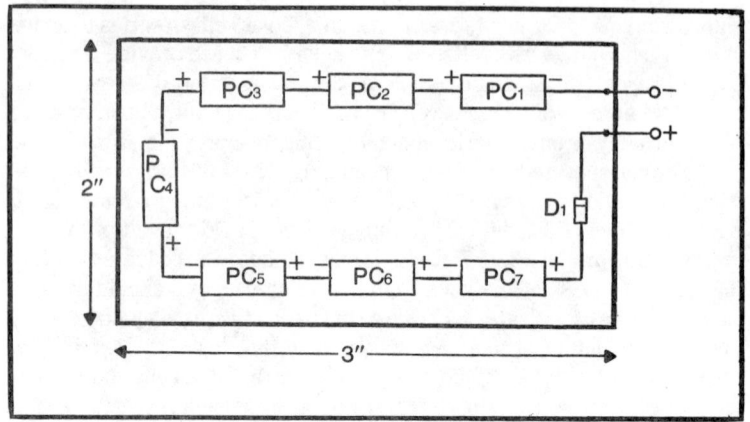

Fig. 4-8. Circuit layout on perforated circuit board.

Figure 4-8 shows the layout of the circuit on a piece of perf board measuring 2 x 3 inches. This is about the smallest size that will accommodate all seven cells and the diode. Make certain that all the cells are correctly wired with the positive terminal of one connecting to the negative terminal of another, and so forth. A reversal of any one cell will cause the entire circuit not to work. The diode must be installed as shown with the unbanded end, or anode, connected to the positive terminal of the solar cell. The cathode end serves as the connection for the positive output lead. A reversal here will completely shut off all current flow from this circuit.

The solar cells may be attached to the board with epoxy cement and all wiring kept on top. An alternate way would be to mount the solar cells on one side and push all component leads through the circuit board holes, wrapping them and making solder connections on the bottom. If this latter method is used, it will be necessary to keep the bottom section of the board away from any metallic objects which might cause a short in the circuit. Once all connections are made, re-check to make certain that every element of this circuit has been connected in a properly polarized manner. Unlike previous circuits, The solar power supply is composed of nothing but polarized elements, i.e., photovoltaic cells and the diode.

Check-Out Procedure. To check the solar power supply, all that is required is a voltmeter which will read up to 3.5 VDC. Preferably, the meter will have a low-voltage scale no higher than ten volts. It is much easier to get an accurate reading of voltage when the value being measured can be read on the center or upper portion of the scale where movement accuracy is best. Place the positive probe of the voltmeter on the positive lead of the power supply. Do the same with the negative probe and lead. Now, expose the power supply to a bright light source, preferably the direct rays of the sun. You should measure a voltage of approximately 3 VDC. Depending on the individual cells, the center value of 3 VDC may be exceeded by a small fraction of a volt. Some cells will deliver slightly less voltage. Small variations should make no difference whatsoever in the operation of the devices they power.

Once it has been established that the circuit is operating properly, it may be connected to any low-current circuit which was formerly powered by a 3-volt battery supply. To determine the current drain of a device, connect it to its battery power source through a milliammeter. Most multimeters will take current measurements of this type. Figure 4-9 shows how this is done. The negative input to the circuit is connected directly to the negative battery lead, while the positive input is connected through the meter. The positive meter probe attaches to the positive circuit input and the negative probe is attached to the *positive* output of the power supply. When the circuit is activated, the current drain can be read directly on the meter. If this drain is less than 20 milliamperes, the solar power supply circuit will provide adequate output. If higher than 20 milliamperes, a more powerful supply will have to be used. It is easy to construct a more powerful supply which will deliver many times the current output of this one. Simply substitute the seven cells with seven

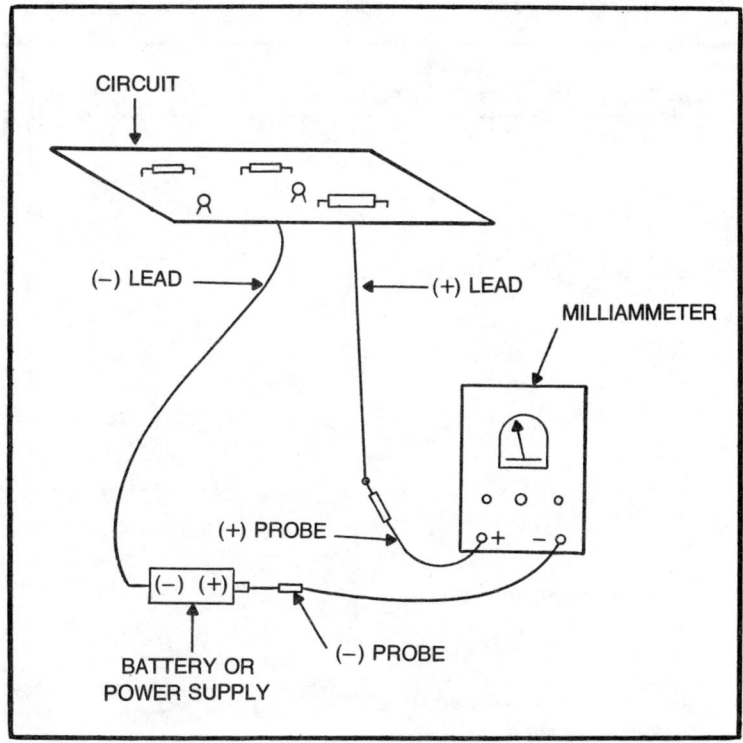

Fig. 4-9. Connection of multimeter to determine current drain of electronic device.

others which have higher current ratings. All else in the circuit will remain the same.

SIX-VOLT HIGH-CURRENT SUPPLY

This project is very similar to the previous solar power supply except it provides a 6 VDC output at a current of 1 ampere in bright sunlight. This is not an inexpensive supply because each solar cell will cost about $10, and 14 of them are used to supply the stated voltage and current.

The solar power supply shown in Fig. 4-10 is a bit fancier than the last one, since it contains a built-in current meter which reads up to 1.5 amperes. There are two SPST switches: one is the main power switch which is used to activate the electronic device under power; the other is shunted across the ammeter. When the shunt switch is in the open position, the current path is through the meter; so an indication is given on the dial face. When the shunt switch is closed, all current is shunted around the meter and no indication is given. The reason for the shunt switch may not be immediately obvious. Since the total output from this power supply will be a maximum of 1 ampere of current under bright sunlight, there is no chance that the meter pointer could be driven off-scale, damaging the

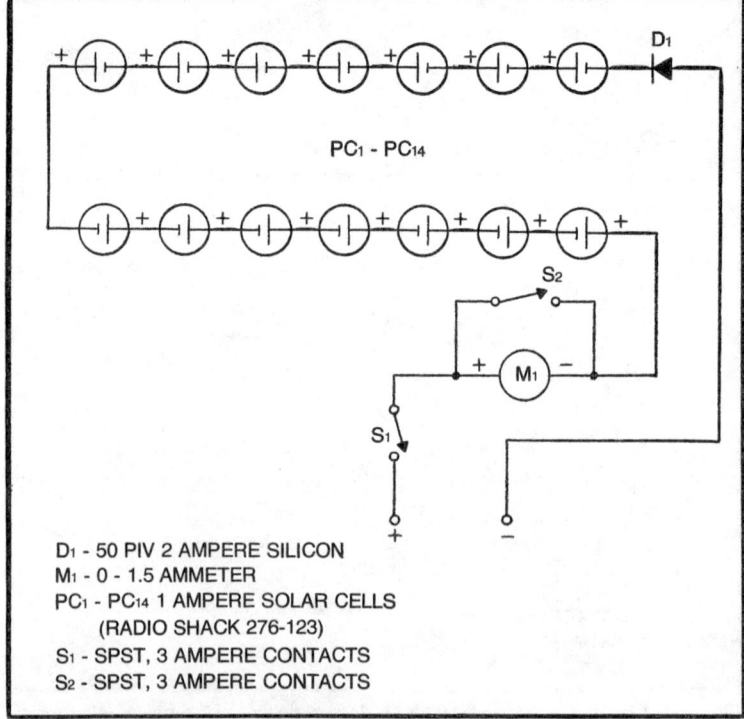

Fig. 4-10. Six-volt solar power supply.

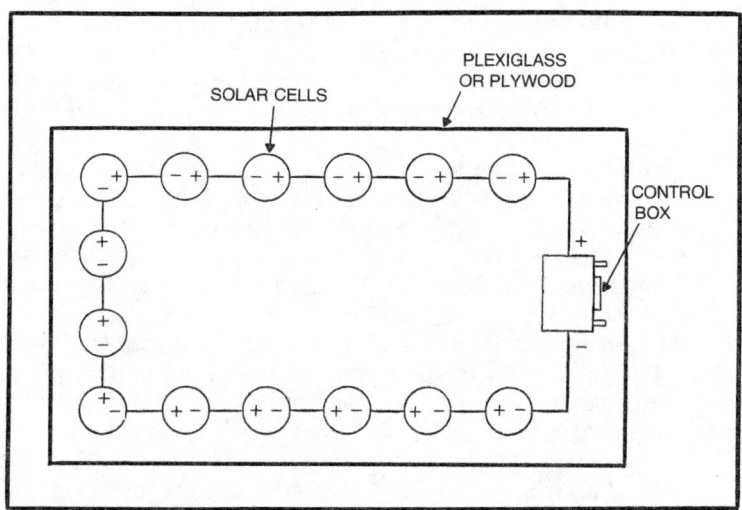

Fig. 4-11. Mounting of circular cells on plywood or plexiglass sheet.

mechanism. No, the switch is not for meter protection purposes; its sole purpose is to avoid the minute voltage drop that the meter's presence in the circuit creates. While an ammeter has very little effect on any circuit, solar cell power supplies are often of such low voltage output that every fraction of a volt counts. Some ammeters may cause slightly higher voltage drops than others. The switch can be open to obtain a reading of current output to the device under power. It is then closed for maximum voltage until another meter reading is desired. Some experimenters may wish to omit S2, which is satisfactory if the voltage output is adequate for the electronic devices to be powered. The other switch, S1, simply makes and breaks the current path between the supply and the electronic device. It is the on-off switch. Both switches are common types and should have current ratings of at least 1.5 amperes. Standard 3-ampere units are specified in this schematic because they are inexpensive and quite easy to find. Any switch rated at 1.5 amperes or more, however, will be adequate.

A protective diode has been placed in this circuit in case the solar power supply is to be used to charge batteries. The diode prevents current from flowing back into the solar cells from the charged battery. If the supply is to be used to directly power electronic devices which do not contain batteries, the diode may be omitted. Without this device, the voltage output will increase slightly to a nominal 6.3 VDC. This minute increase is of no real concern and will be compatible with all devices requiring a 6.0 VDC power supply.

Figure 4-11 shows how these cells might be mounted on a moderate-sized sheet of plywood or plexiglass. Any other fairly firm material will do as well but make certain that it is resistant to temperatures in the 150 degree range. A small box made from aluminum stock is also mounted on the base containing the solar cells. Here is where the meter, diode, and switches are mounted. This is more practical than leaving them exposed to the elements. Figure 4-12 shows the interior wiring of this box, which is quite simple. D1 is mounted to the output terminal by means of a tie-down terminal strip (Radio Shack #274-688), as is shown in the inset. The unused sections of this strip may be clipped away with a pair of diagonal cutters. Notice that the diode is connected in the negative leg of the power supply and its polarity is reversed accordingly when compared to the diode connection in the positive leg of the previous power supply project.

Check-Out Procedure. After making certain that all wiring is correct and that the solar cells are connected with all polarities aligned as in the schematic diagram, turn switch S1 to the closed position. Do the same with S2. Connect the probes of a voltmeter to the positive and negative output terminals located through the aluminum box. Expose the solar cells directly to the sun's rays. Since this is a high-powered supply, only the direct rays of the sun will be able to provide full output. You should immediately get an indication of about 6 volts on the external voltmeter. If there is no indication, re-check S1 to make certain that it is actually closed and completing the circuit. If S1 is in the proper position, place the voltmeter probe directly across the solar cells at the first and last units in the string. The negative probe should be placed across the negative terminal of the solar cell to which D1 is attached. The positive probe is placed across the positive contact of the solar cell which is connected to M1. If there is no reading here, you have a defective solar cell, broken wire in the solar cell string, or a cell which has been reverse-connected in the string.

If voltage is present at the output of the solar cell string, keep the positive probe in position and place the negative probe on the output side of the diode. If voltage is not present here, the diode is defective or has been reverse-connected in the circuit. If voltage is present on this side of the diode, the problem will be found in the form of a broken wire between these points and the final output terminals.

During the initial check-out, if voltage is present at the output terminal, open S2. This is the switch across the meter terminals. With the external voltmeter in place across the output terminals, a reading should be obtained with the switch in the open position. If the reading drops to zero when S2 is open, this means the meter is defective and must be replaced. You will not get any meter indication during this check-out procedure, because almost no current is being drawn by the voltmeter circuit.

To further check power supply operation, connect an electronic or electrical load to the output terminals. Be sure to observe polarity. Preferably, this load will draw a current of around 500 milliamperes or more, but less than 1 ampere. The exact amount of current drain should be determined ahead of time using a conventional power supply and milliammeter.

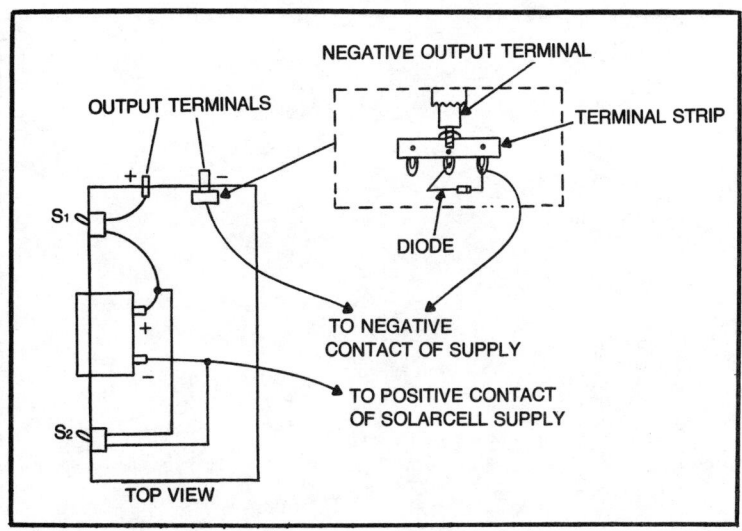

Fig. 4-12. Interior wiring or termination and metering box.

With the load connected, activate the power supply by closing S1, opening S2 and noting the reading on the ammeter. It should be close to the known current drain of the device under power. Slight deviations between two different types of meters can be expected and should not be of concern. If it is anticipated that currents of less than 500 milliamperes will be drawn by most devices powered by this supply, M1 may be replaced with a lower-reading unit.

For those of you who require a 12-volt power supply with similar current output, this circuit can be modified by doubling the amount of cells in the series string. No other components will have to be changed, and the output will be a full 12VDC. The cost of the supply will nearly double. The 6-volt version should cost around $150.00, with $140.00 of that representing the cost of 14 solar cells. A 12-volt unit would cost about $290.00; again, at $10.00 a piece, the solar cells are always the most expensive item.

VERSATILE LIGHT-CONTROLLED SWITCH

Day-night switches were briefly discussed in an earlier chapter. Figure 4-13 shows a schematic of a very versatile switch which has two outputs for 115 VAC lights or appliances. The electronic circuit is rather conventional, consisting of an npn transistor, miniature solar cell, variable resistor, miniature relay, and 9-volt battery. The relay chosen for this project is a miniature version designed to operate from 9 volts DC and will close when 12 milliamperes of current flow through its windings. Its contacts are rated at 1 ampere, so only low-powered devices may be plugged into the receptacles provided.

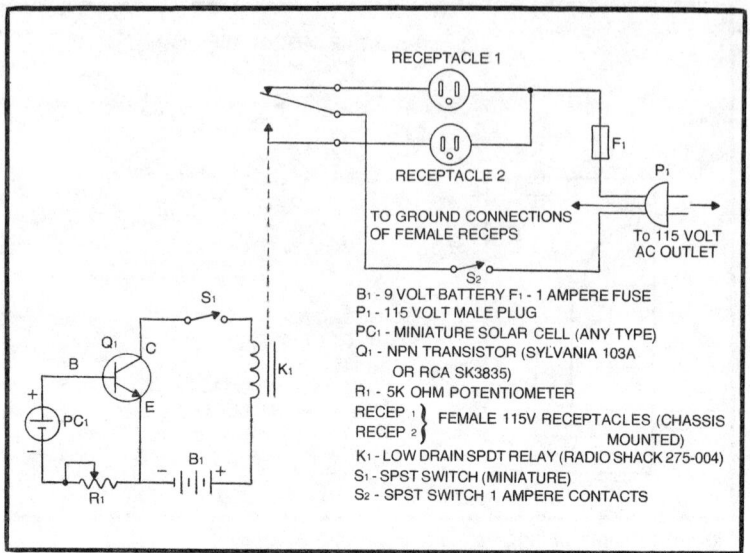

Fig. 4-13. Light-controlled switch circuit.

Figure 4-14 shows parts-placement diagrams of the perf board and the metal or plastic case which contains the circuit board and receptacles. Begin by wiring the circuit on the perf board. The battery is connected by means of a standard 9-volt terminal clip and is held in place by a metal bracket. Three small pieces of insulated hookup wire are connected between the relay output on the board and the receptacles. The potentiometer and solar cell are connected as shown. Be very careful with the 115-volt wiring because a short circuit here could easily blow a fuse and wreak havoc with your circuit board. A 1-ampere fuse is placed in the 110-volt line to avoid any fire hazards should a short circuit exist. This fuse will also blow should a load be placed on the AC line which draws more than 1 ampere of current, the maximum rating of the relay contacts. S1, which is connected between the relay and transistor, is used to deactivate the electronic circuit; while S2 is used to switch off the 110-volt circuit. With S1 in the off position, the relay will never be activated; but the normally closed set of contacts will be engaged and any device connected to receptacle number 1 will be activated, assuming that P1 is connected to a 115-volt source. With S2 in the off position and S1 closed, the relay will continue to work; but no power will be fed to any devices connected to the receptacles.

Check-Out Procedure. Turn S1 and S2 to their OFF positions. Connect two 115-volt, 15-watt light bulbs to receptacles 1 and 2. Throw S2 to the on position. The light bulb attached to receptacle 1 should light. If it does not, check P1 to make sure it is properly inserted into its 115-volt outlet. F1 should also be checked to make certain it has not blown. If all is well at this point, throw S1 to the on position and direct the beam of a flashlight onto the surface of PC1. The light attached to receptacle 2 may be immediately activated; if not, adjust R1 until receptacle 2 provides power to the light. If no adjustment of R1 will activate receptacle 2, check B1 to make certain that it is fresh and supplying power to the circuit. Listen as you adjust R1 to detect the clicks which will be made by the relay as it engages and disengages. If you do not hear these clicks, check the wiring of the electronic circuit. Pay special attention to the polarity connections of B1 and the solar cell.

If the relay clicks are heard and receptacle 2 still refuses to activate the light bulb, there is a wiring error somewhere in the 115-volt circuit. This could be due to a bad solder connection at the relay contact point or a broken wire somewhere in the circuit.

Remember, never connect any load to the 115-volt circuit which will draw more than 1 ampere of current. This limits you to 100-watt bulbs, although two may be connected, one to receptacle 1 and the other to receptacle 2. Only one light bulb can be activated at a time. If you desire to control higher current loads with this type of circuit, this is very easily done with the circuit shown in Fig. 4-15. This could be called an *extension relay circuit,* and two of them may be built and powered from receptacles 1

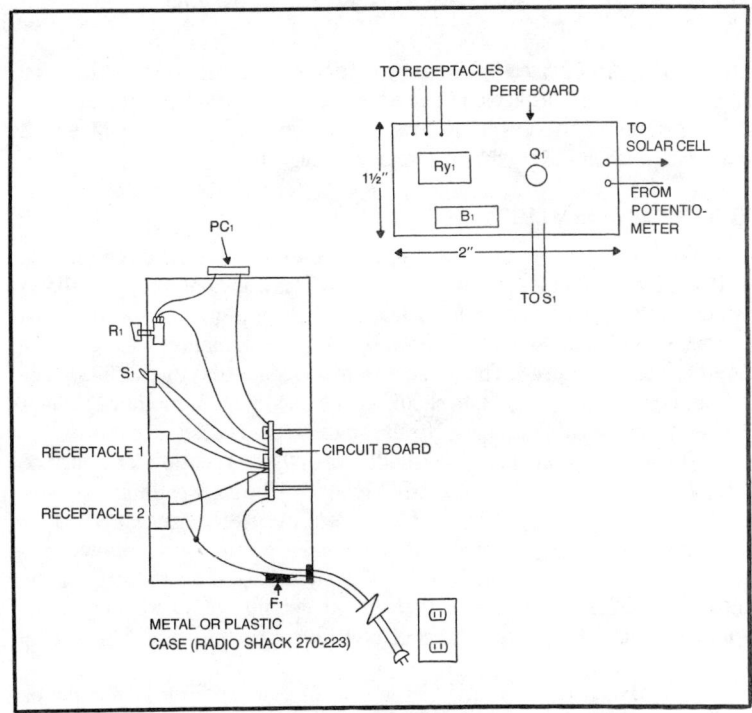

Fig. 4-14. Component placement diagram for light-controlled switch.

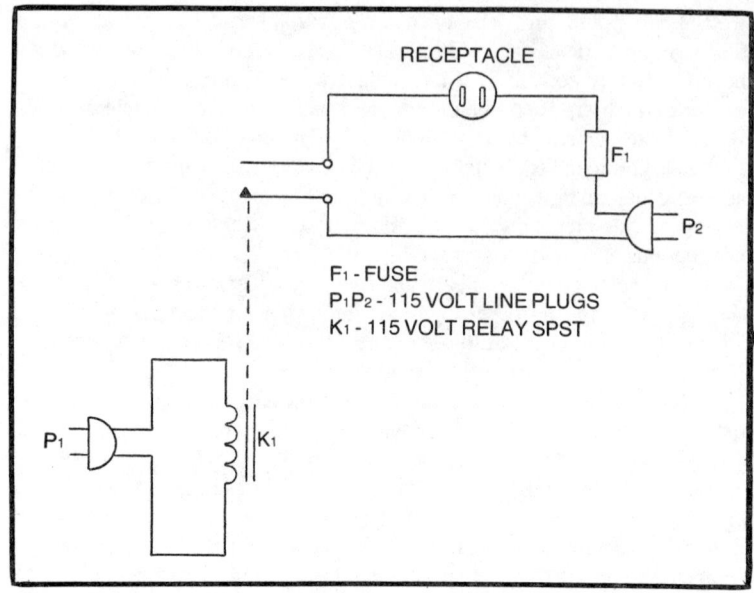

Fig. 4-15. Addition of extra relay increases current-handling capabilities of switch.

and 2 of the former circuit. The relay type is not specified here, although it is of the 115-volt variety and should contain contacts rated to carry the amount of current necessary to power whatever loads are to be connected. The relay coil should not draw more than 1 ampere.

ELECTRONIC ALARM CLOCK

For those readers who like to get up with the sun, the circuit in Fig. 4-16 may provide a pleasant way of doing so. It is a solar powered oscillator which emits a pleasant warbling sound much like the singing of a bird when adequate light is present to power the circuit. Six miniature solar cells are used as a power source. These can be purchased for less than $10.00. The remaining circuit components should cost a maximum of another seven or eight dollars if all are purchased from standard retail outlets.

This is a complete unit and is built on a 3 x 8-inch section of perforated circuit board material, as is shown in Fig. 4-17. The solar cells and speaker are all mounted atop this board. There is really no need to encase this unit, as it may be permanently mounted to the inside portion of a window frame to catch the first light of the sun. S1 allows you to deactivate the alarm on those mornings when you don't have to get up early. R3 allows for adjustment of the sound which is emitted from the miniature 8-ohm speaker.

In building this circuit, the major electronic portion should be assembled in the center of the circuit board first. The parts placement

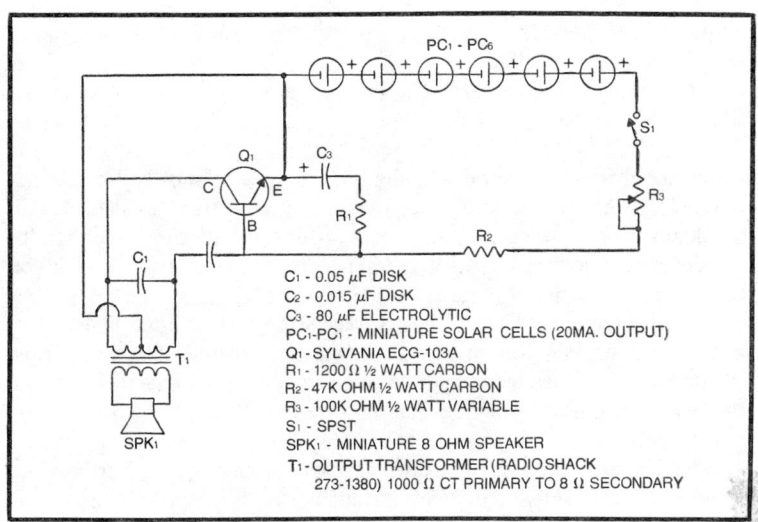

Fig. 4-16. Electronic alarm clock schematic.

layout is shown in Fig. 4-17. Once this portion of the circuit is complete, the solar cells are placed on the right-hand portion of the board and may be held in place with epoxy cement. The miniature speaker is installed on the left-hand side of the board. The diagram shows it mounted in a perpendicular configuration with the board; however, it may be mounted flush with the board by drilling a circular hole to accommodate the speaker magnet on the back.

The polarized devices in this circuit include the 6 solar cells, the transistor, and capacitor C3. Make certain these components are wired as

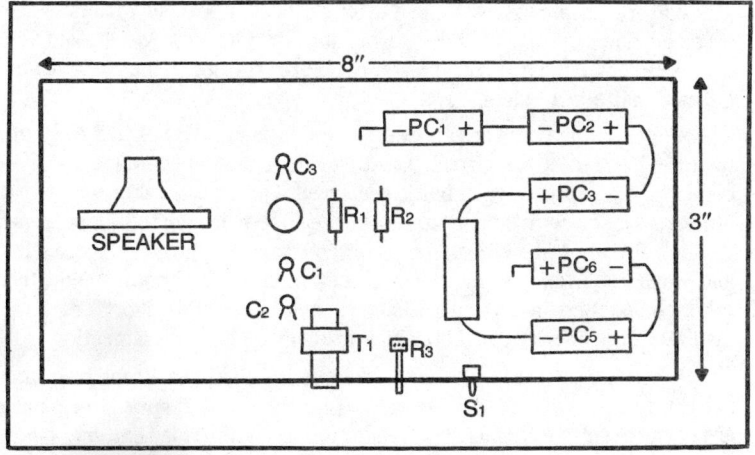

Fig. 4-17. Component placement for electronic alarm clock.

shown in the schematic diagram because reversal of any of them will cause this circuit to not operate.

Check-Out Procedure. This is a very simple circuit to test, but do not turn S1 to the on position until you have made certain that all wiring is correct. If all looks well, activate the circuit by throwing S1 to the on position and direct a bright light at the surface of the solar cells. You should immediately hear a warbling sound from the speaker. If not, adjust R3 until the desired tones are heard. Inoperation of this circuit usually results from a wiring error or components wired in reversed polarity. Check for all of these conditions should you not obtain the desired results.

This circuit will have to be adjusted from time to time due to the changing natural light intensities of the various seasons. Obviously, this circuit cannot be depended upon to wake you up at the exact same time every morning because of the variations in light intensity brought about by clouds, rain, and other weather conditions. It is an interesting circuit to assemble and experiment with, however, and can be operated from normal room light, a flashlight, or even the light from a campfire. For nature outings and bird watchers, this project may be a natural.

LIGHT-CONTROLLED ELECTRONIC ORGAN

For the musicians in the readership, a light-controlled musical instrument can be a fascinating project to assemble and use for years to come. This one is powered from a 6-volt lantern battery and offers a reasonably loud output signal which will last for many hours before the battery is discharged. Shown in Fig. 4-18, this musical organ is played by moving your hand up and down and back and forth over a cadmium sulfide cell, or photoresistor.

This is an unusual circuit because it is mounted atop the lantern battery. Two holes are drilled in the circuit board, as shown in Fig. 4-19. The battery terminals are slipped through these holes and wire nuts are screwed down on top of these terminals. Two hookup wires from the center tap of T1 and from the emitter of Q1 are attached to the positive and negative battery contacts.

Begin building this circuit by drilling the required holes in the circuit board. The size of the circuit board will depend upon the size speaker used, with miniature types being preferred. Obtain a speaker which has brackets for screw-in mounting. Bolts are slipped through these brackets, through slightly widened holes in the circuit board, and attached with nuts and washers. After the speaker has been fixed to the board, mount the other components, as shown in Fig. 4-19, and solder. Two short lengths of insulated hookup wire with bared ends are used for battery connections.

When the circuit is complete, remove the protective caps from the battery terminals. Lantern batteries which have coiled spring terminals are not as desirable as those which offer the small, threaded battery posts. The latter can be easily slipped through the drilled holes in the circuit

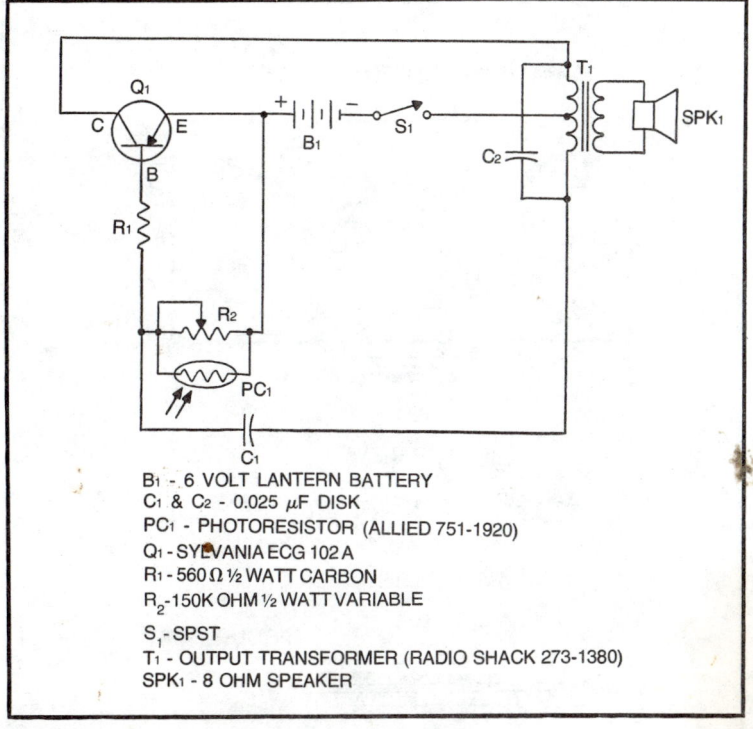

Fig. 4-18. Light-controlled electronic organ circuit.

board and the wire nuts screwed back down, securing the hookup wire and circuit board to these posts. When it is necessary to replace the battery, the wire nuts are removed, the battery slipped free, and a new one inserted.

After the circuit board and hookup wires have been attached, make certain that all contacts between the circuit and power supply are firm and that no short circuits have occurred through the procedure of mounting the board atop the battery terminals.

If all appears well at this point, activate the circuit by throwing S1 to the on position. The circuit is complete if a tone is heard in the miniature output speaker.

Check-Out Procedure. When S1 is thrown to the on position, a tone should be heard immediately. If this is not the case, check the battery with a voltmeter whose probes have been placed at the positive and negative connections. If the voltage reading appears normal, a wiring error or faulty component is probably the cause. Make certain the transistor has been properly installed and that the positive and negative connections of the battery are attached to the correct circuit points. An error with either the transistor or battery connections can result in the transistor being

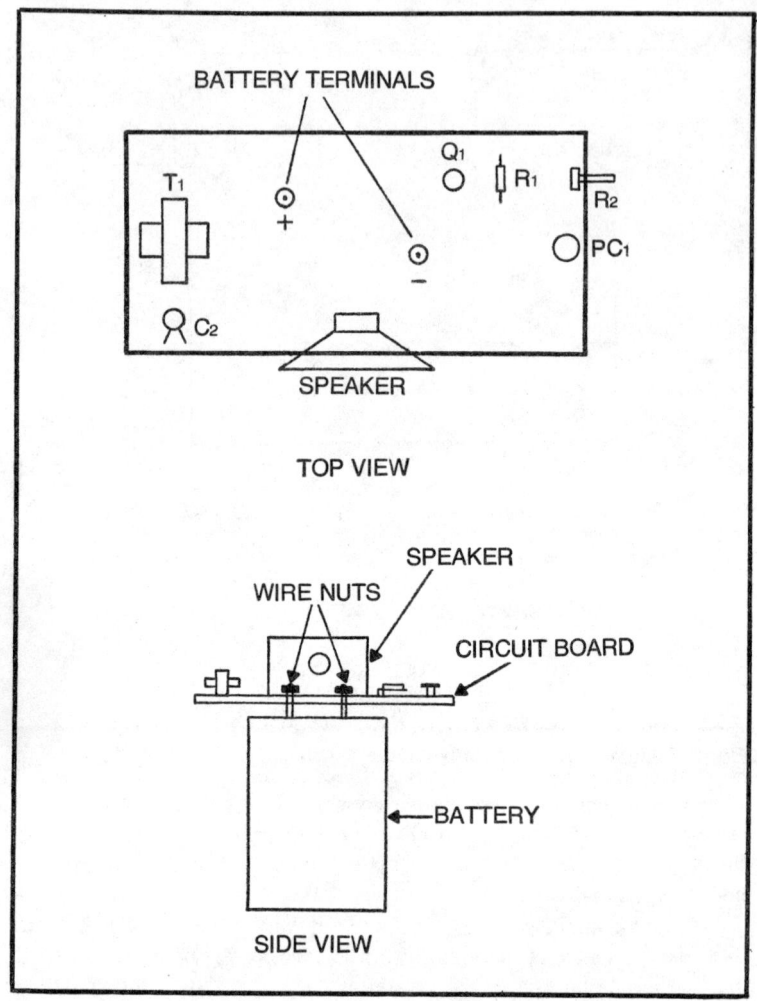

Fig. 4-19. Placement of components for electronic organ on circuit board.

permanently damaged. If you discover a wrong connection here, correct the situation and test the circuit again. If a tone is heard, no permanent damage has occurred. If a tone is still not heard, check the wiring of the remainder of the circuit. If no problems are found here, the former misconnection may have damaged the transistor and it should be checked or replaced.

Overall frequency adjustment can be set by variable control R2. Once the range has been established by this device, a fairly strong light source should be played upon the treated surface of the cadmium sulfide photoresistor. Now, by moving your hand back and forth between the light and the photocell, you will be able to vary the output frequency at a controlled rate. With a little practice, you should be able to play simple songs through hand movement alone. With more practice, you may be able to use one hand to vary the light intensity to the photoresistor while your other hand controls R2. Using the two-handed approach, a wide range of audio frequencies will be heard at the speaker output.

AM RADIO BOOSTER

Many inexpensive AM radios of "shirtpocket" size work well during the nighttime hours when static created by the sun is not a factor. But during the day, reception may be poor for all but the closest local stations. Figure 4-20 is a preamplifier circuit which is designed to boost the sensitivity of any AM radio. It works directly from sunlight and may be attached to any type of AM radio. This circuit is especially adaptable to crystal radios which use no other power source. It can also be used with some of the solar powered AM radio circuits which will be discussed in this book.

This circuit works best if its input can be connected to a length of wire which serves as an antenna. The longer and higher this wire is, the better the improvement in radio reception. The output from this amplifier is connected directly to the antenna terminal of your AM radio. If this type of terminal has not been brought to the outside of the radio case, you will have to remove the case and make a solder connection directly to a winding of the ferrite-rod antenna.

Figure 4-21 shows a circuit board layout of this circuit which uses three rectangular solar cells with an output of about 20 milliamperes each.

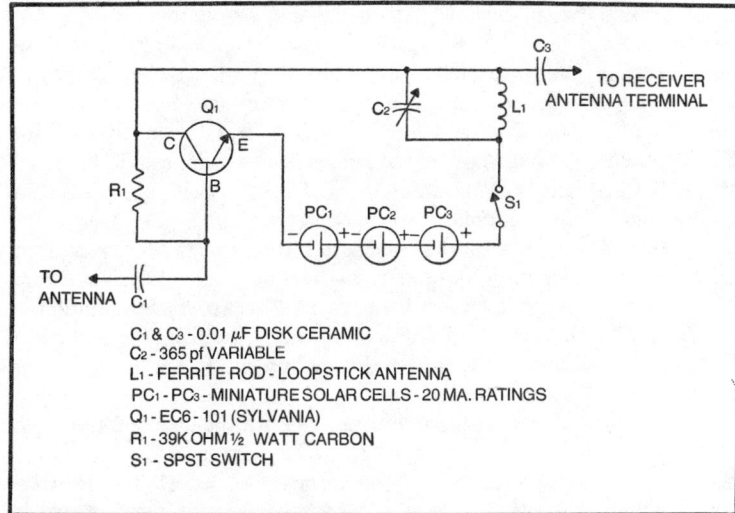

Fig. 4-20. AM radio booster circuit.

A separate antenna coil and tuning capacitor will allow you to peak the amplifier to the received frequency by tuning a weak station on your AM radio and then varying the position of the rotor in C2 until maximum signal strength is obtained. This circuit can be mounted in a small plastic box or may even be attached to the top of a small transistor radio. It may also be built into a plastic cabinet containing a home-built crystal-or solar-powered AM radio. Due to the number of ways this solar-powered preamplifier may be used, no specific enclosure mounting instructions are provided here.

Referring to Fig. 4-21, attach the antenna coil, L1 to the left edge of the circuit board using epoxy cement. C2 may be connected in the same manner. C2 is a variable capacitor and you must be very careful not to let any epoxy drip on the rotor portion, or this component may be ruined. Most variable capacitors are equipped with brackets which enable them to be mounted to circuit board with nuts and bolts.

Once the two major components have been installed, insert the remaining capacitors, resistor, and transistor in the circuit board at the positions shown and connect them according to the schematic drawing. Solder these connections.

The three solar cells are now attached to the right-hand portion of the board. These may depend upon their wire leads pressed firmly through the circuit board for secure mounting or they may be epoxied to the board. Make certain the three are series-connected, as shown in the pictorial drawing. Before connecting the leads from the solar cells to the remainder of the circuit, make sure that S1 is in the off position. S1 may be mounted through the circuit board or may be epoxied to one edge, as is shown in Fig. 4-21.

Connect the antenna lead to a suitable length of wire for AM reception. A single length of wire 60 feet long suspended from the roof of a home or top of a tree is ideal, although much shorter lengths will work too. Connect the other wire to the receiver antenna terminal. Now, expose the solar cell to bright sunlight and tune your AM radio to a weak station. Turn S1 to the on position. You may notice an immediate increase in signal strength. Rotate the shaft of C2 until the signal peaks, as is indicated by the loudest signal in the AM radio.

Check-Out Procedure. If the above steps do not produce increased signal strength, you must look the circuit over once again to make certain all wiring is correct. If it appears to be, check Q1 to see if it is operational. This can be done with a transistor checker, or it may be returned to its place of purchase if you do not have this instrument. If there is any doubt about the transistor, replace it.

The length of hookup wire between the preamplifier and the radio receiver should be as short as possible. If this wire is unduly long, your problem could be occurring here. Make certain, too, that S1 is really in the "on" position. You could have attached it to the circuit board in such a manner that you are confused as to which direction is on and which is off.

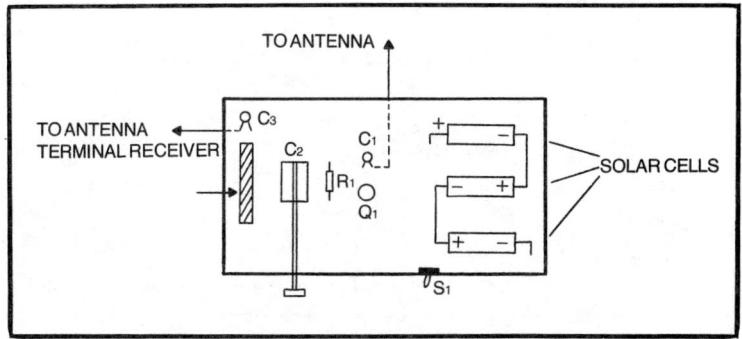

Fig. 4-21. Booster circuit board component layout.

If you still get no results, measure the output voltage from the solar cells with a voltmeter. You should get a reading of just over 1 volt (nominally, 1.35 VDC). If there is no voltage at this point, you may have accidentally reversed one of the three solar cells resulting in a zero voltage output. If normal voltage is obtained from the cells, check to make certain that the positive lead is connected to the proper portion of the circuit. Sometimes these supplies are easily reverse-connected so that the negative lead is attached to the positive circuit contact and the positive lead attached to the negative contact. In this circuit, fortunately, a reversal of the power supply should not damage the transistor or any other components. Correctly connecting the solar power supply should alleviate any problems and normal circuit operation will result.

SOLAR POWERED FM RADIO

Many of us have experimented with crystal AM radio sets and miniature transistorized receivers designed to receive the AM broadcast frequency. Some of the projects in this book deal with just these types of circuits. Rarely, though, does one see a circuit for a miniaturized radio designed for reception on the FM broadcast band. The circuit shown in Fig. 4-22 is designed to cover the entire FM broadcast band; and the receiver will even tune in airport frequencies, as well as the entire two-meter amateur radio band. Admittedly, its small size and circuit simplicity combine to make this a rather insensitive device, but local broadcasts will be plainly heard while using only a 14-inch whip antenna.

The entire circuit is powered from the sun or any other convenient light source. Two transistor amplifier stages are employed to boost the output to a level which will drive an electronic earphone with adequate volume. This receiver is really a crystal detector much like the powerless crystal radios mentioned before. After the detection stage, however, a sun-powered amplification circuit boosts the miniscule output to a level which may be comfortably heard through the earphones. Due to this circuitry, the FM receiver is far more sensitive than a standard crystal set.

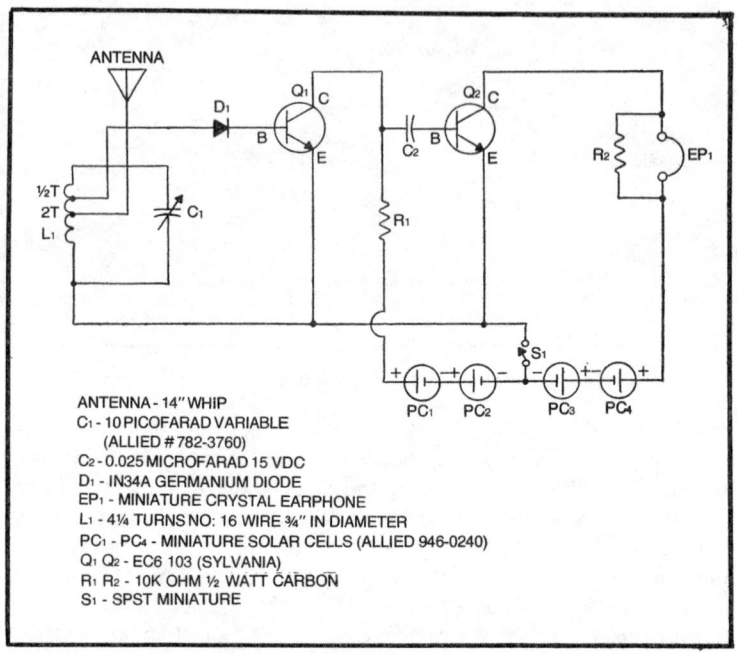

Fig. 4-22. Solar powered FM radio circuit.

Figure 4-23 shows the circuit board layout. L1 is wound by hand and consists of 4½ turns of #16 tinned copper wire spaced evenly over a distance of ⅝ inch. The diameter of the coil is ¾ inch. Figure 4-24 shows a closeup view of the dimensions of this component. Wind this part on a cylindrical object such as a pencil or screwdriver shaft. The winding form must be slightly less than ¾ inch in diameter. Once the coil is removed from the form, it should measure ¾ inch across the center of one turn. Wind this component as closely to specified dimensions as possible. Later adjustment can be made by squeezing the turns of the coil a little closer together or spreading them apart. When you first wind this coil, all of the turns may be touching each other. The coil is removed from the form and the point of a pencil can be used to spread each winding until the total coil length measures ⅝ inch. When completed, no coil turn should touch another.

Place the electronic components on the circuit board, as shown in Fig. 4-23. Adhere pretty closely to the pictorial layout of Fig. 4-22 for parts placement. Try to keep the coil away from any nearby metallic objects. C1 is a variable capacitor known as a "padder". This is adjusted by turning the center-mounted bolt with a small screwdriver. Ideally, the screwdriver will be plastic since a standard metal tool will cause the frequency to shift as it gets close to the coil and capacitor. The four rectangular solar cells are mounted to the bottom right-hand portion of the board and connected as

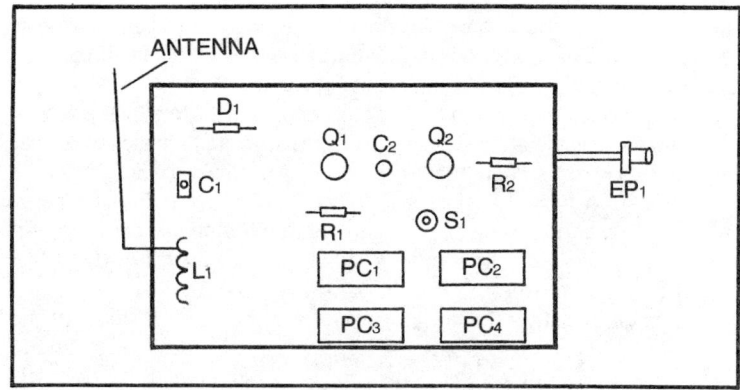

Fig. 4-23. FM radio circuit board layout.

shown in the schematic. Caution: these cells are connected differently than with all the previous circuits. The four cells form two series circuits with two cells in each leg. Refer again to the schematic and note that PC2 and PC3 both have their negative leads attached to each other and to S1. Also, make certain D1 connected as shown with its cathode attached to the base of Q1.

Make certain S1 is in the off position before making the final connections from the solar cell bank to the circuit. When all wiring is complete, closely examine all connections. This circuit is a bit more sensitive than some of the others and will require strict adherence to sound building practices. Notice that the antenna is attached at the second turn of L1, while the lead to the diode is attached at one-half turn from the same end. These connections are most important, so count the turns properly. A 14-inch length of stiff hookup wire can serve as an antenna.

Check-Out Procedure. After checking all wiring one more time, you are now ready to check the operation of your circuit. For this first test,

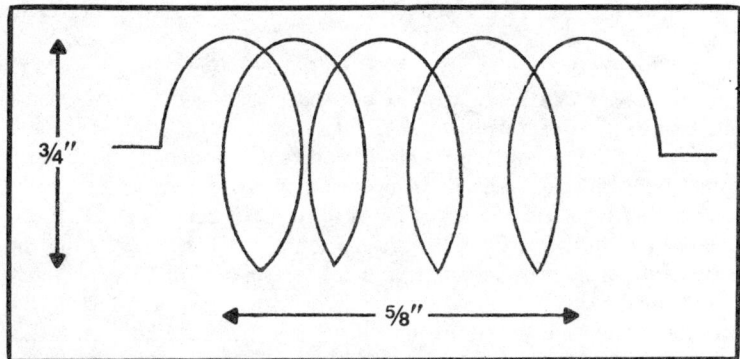

Fig. 4-24. Closeup view of FM radio coil dimensions.

105

if you can attach the 14-inch antenna wire to a beam-type of FM antenna used with stereo receivers, check-out will go more smoothly. If a beam antenna is not available, however, setting up this circuit is still a fairly simple procedure.

With the antenna set in the best location, turn S1 to the on position. You should hear a faint "rush" of noise in the earphone. You may even hear an FM broadcast station. With a plastic screwdriver, adjust C1 until a signal is strongly heard. Now, with a commercial FM radio, attempt to tune in the same station and determine where it lies on the dial. Once you have aligned the two radio receivers, slowly tune the commercial receiver toward the upper or lower end of the FM broadcast band. Just move a little at a time until another strong station is heard. Then attempt to tune your home-built receiver to the same strong frequency. The low end of the FM broadcast band is preferred for this test. If your home-built receiver will not tune all the way to the low end, squeeze the antenna coil, L1, very slightly to compress the windings. This will shift the entire frequency range of your receiver, causing it to tune lower. You have completed receiver alignment when your project will tune just to the low end of the FM broadcast band. Now, by reversing the direction of the capacitor control bolt, you can tune in the opposite direction to the top of the FM band and proceed from there to tune as high as about 150 megahertz. This tuning range will vary and more squeezing or separation of the coil will be necessary. For instance, if a station at the low end of the FM broadcast band were tuned by the home-built receiver in the middle of the capacitor adjustment range, the coil would be separated a bit to cause the frequency range to shift higher. Your goal is to match the lower end of the tuning range of your receiver to the lower end of the FM broadcast band.

If your circuit does not operate, you may have reverse-connected one or more of the solar cells. A sure sign of a wiring error is the absence of any sound at all in the earphone. Also, check for a reverse-connected transistor or diode. In weak signal areas, it may be desirable to obtain the loan of an rf signal generator which will tune the 88 to 150 MHz range. The output from this generator can be connected directly to the antenna terminal. This will assure you of having a strong signal at the receiver input.

After check-out is complete, the entire circuit may be mounted in a plastic case. A metal case may also be used, providing that the antenna is inserting a plastic screwdriver. Alternately, a miniature variable capacitor tuning, and the coil may have to be compressed or widened a bit more. A small hole drilled in the plastic or aluminum case over the mounting area of the tuning capacitor will allow for easy receiver tuning by inserting a plastic screwdriver. Alternately, a miniature variable capacitor may be obtained which may be tuned with a knob instead of a screwdriver. Miniature variables are sometimes difficult to come by, which is the reason for choosing the more common type specified in this project. Either will work equally well with the conventional type offering more convenient tuning.

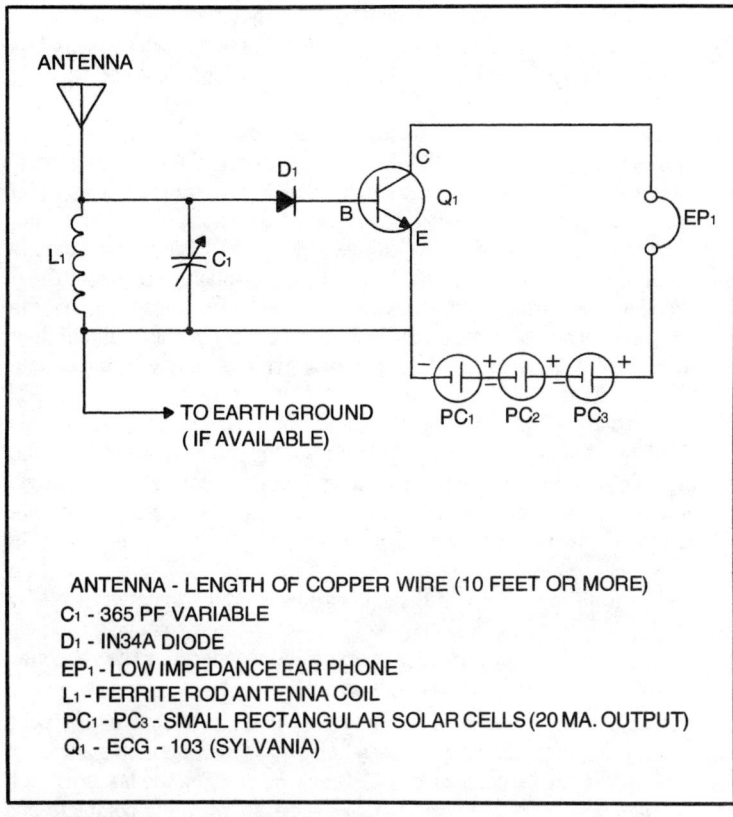

Fig. 4-25. Solar powered AM radio circuit.

SOLAR-POWERED AM RADIO

Having discussed a project which involved the reception of FM broadcasts, it is only fair to also offer a circuit which is more conventional, in that it tunes the AM broadcast band. The circuit shown in Fig. 4-25 is very similar to the FM receiver, although the coil and capacitor are different because of the different frequency range. Also, there is only one stage of amplification. This circuit is especially designed to be used in conjunction with an earlier project, the AM broadcast radio booster. As a matter of fact, this project uses the same power supply as the booster. Because of this, a separate power supply is not necessary for the AM radio if the booster circuit is also to be used. Referring again to Fig. 4-25, the emitter and earphone connections to the indicated solar cells may be made to the solar cells in the booster project. Both of these projects could even be built on one circuit board, and the antenna of the AM radio eliminated and replaced by the receiver antenna connection from the broadcast booster.

Assembly is straightforward and can be handled in the same manner as with the FM receiver project. L1 is a ferrite antenna which is sold by many different manufacturers. It consists of a powdered iron rod which is wound with many turns of insulated wire. The tuning capacitor is a broadcast variable type which is also available at most hobby outlets.

Referring to Fig. 4-26, it can be seen that the ferrite antenna is mounted to the extreme left of the small circuit board. It is held in position by epoxy cement. The tuning capacitor is positioned near the coil, although it may be placed closer to the edge of the circuit board than is shown in the drawing if a long-shaft model is not available. D1 should also be placed in close proximity to the coil. Install all of the major components first, making all solder connections before mounting the solar cells, if they are to be used. Connect the cells in a series string as shown, being careful not to reverse any of the leads.

A word about antennas might be in order at this point, because this type of radio depends heavily upon this item for receiver sensitivity and volume. The antenna should be at least ten feet long, consisting of a single section of flexible copper wire. Greater lengths are even more desirable. All antennas should be mounted in a vertical position, if possible, in relationship to the ground. This type of receiver can boast good reception of local radio stations. Even distant stations may be received well by connecting the ground lead to a cold water pipe within your home's plumbing system. A good earth ground can do wonders for the receiving efficiency of this type of antenna system.

Check-Out Procedure. No on-off switch has been provided with this circuit, although one may be installed between the positive output of PC3 and the earphone connection. Between connecting the last solar cell to the main portion of the circuit, make sure all wiring is complete and accurate according to the schematic. Connect the final cell and listen through the earphone. The earphone must be the low impedance variety or magnetic type. Crystal earphones will not function properly in this low-powered circuit. You should immediately hear a rush of noise. Tune the variable capacitor until a local station is received. Moving the antenna to a better location may improve the volume.

If nothing is heard, check the solar cell connections and make certain they are receiving ample light from the sun or another bright source. Diode D1 may also be suspect, as these devices are often reverse-connected in electronic circuits. If D1, Q1 and the solar cells are correctly connected, your circuit should work. If not, the diode or transistor may be defective; but check the output voltage from the solar cells with a voltmeter to make certain they are operational. An output of just over one VDC should be recorded. If this reading is zero or just a fraction of a volt, re-check all connections. Another cause could be a lack of adequate light striking the surface of cells, or even a defective cell in the string. Replacement of any damaged components will assure a working AM radio powered from the sun.

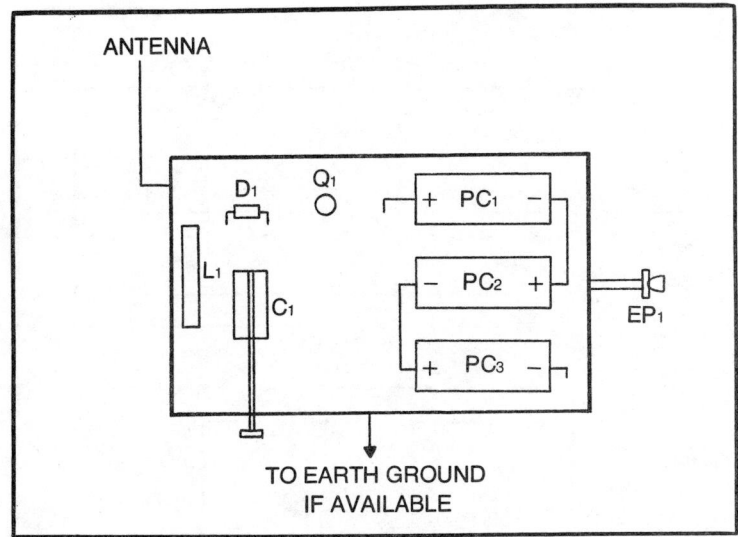

Fig. 4-26. Parts placement for solar powered AM radio.

12-VOLT BATTERY CHARGER FOR HALF THE PRICE

One of the major problems with solar cell experimentation is the cost of individual cells and their low output voltage. Even to make a simple 1.5 volt supply, at least three cells are necessary, and preferably four if you are going to use a diode in the circuit. If these three cells are to be high current units, the price of an individual cell will most likely be around $10. This would mean that the solar equivalent of a "C" cell battery would cost in excess of $30. Of course, it would never run down.

You will notice that most of the projects in this chapter are powered by solar cell strings with an output of six volts or less. Of course, the 6-volt units use about fourteen cells and can be very costly even when moderate current output is desired. Twelve volts dc is pretty much an electronics industry standard for powering many types of circuits. Although 9-volt supplies are also quite common, 12 VDC is the most often used quantity.

To build a solar power supply with an output of 12 VDC would require a minimum of about 28 cells. If the output from the supply were to be 1 ampere, the total cost would exceed $280 using commonly available parts. Due to this cost factor, whenever lower voltages can be used they generally are. There is a way around this cost factor for 12 VDC supplies, and Fig. 4-27 shows one way.

In this schematic, you will see that the solar circuit is a battery charger supplying current to two 6-volt, rechargeable cells. These batteries are connected in series so that their total output is 12 VDC but they are individually charged, as single 6-volt batteries, from a 6-volt solar cell supply. A double-pole/double-throw (DPDT) switch is manually

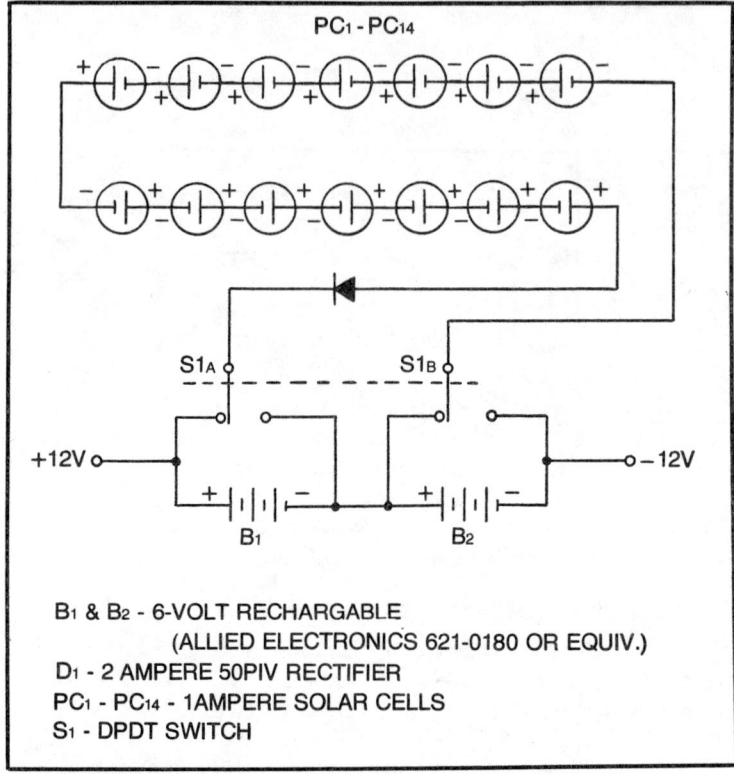

Fig. 4-27. Twelve-volt battery charger circuit.

thrown from time to time to see that both batteries receive approximately equal charges. This must be done manually because only one battery is charged at a time.

The cells specified for this project are 1-ampere models because the batteries are of a size which will take this amount of charging. If smaller batteries were used or if you did not need to charge at the 1-ampere rate, miniature solar cells could be used. As a matter of fact, any type of solar cell would be acceptable and would be determined by the amount of charging current needed for the battery based on current drain from the equipment under power. Diode D1 is placed in the circuit to prevent damage to the solar cells from overcharged batteries.

Figure 4-28 shows the connection of the two batteries to each other and to the DPDT switch and solar cell supply. Get these connections right or the power supply will not operate correctly into the batteries. A reversal of polarity could even occur and damage the cells in one or both batteries. Use a switch which is rated for the current which will be provided by the solar supply.

You must remember to switch the power supply to a different battery every hour or so to provide an even charge. If sky conditions are such that the intensity of the sun changes frequently, you may want to throw the switch to the opposite charging position every half hour or so. This is not really a big problem but requires attended operation. The supply cannot be left all day in one position unless it is left in the opposite position for all of the following day. Only in this manner will an even charge be placed in both cells.

No specific mounting instructions for the solar cells have been provided. The particular cells chosen will determine what form the mounting should take. Small solar cells might be installed on a large section of perforated circuit board, while the 1-ampere cells specified will require mounting on a large sheet of plexiglass or other insulated material. You may power electronic circuits from this supply while the batteries are charging without fear of damage.

Check-Out Procedure. Wire the circuit as shown but before connecting the leads from the switch to the battery terminal, correctly identify them with an ohmmeter. With the switch in one position and the solar cell bank in sunlight, place the probes of the ohmmeter across the two terminals lying parallel and in the position to which the switch lever is pointing. With the positive probe on the positive contact and the other on the negative contact, you should get a reading of a little over 6 VDC. If, on the other hand, the meter needle dips below the zero reading on the scale,

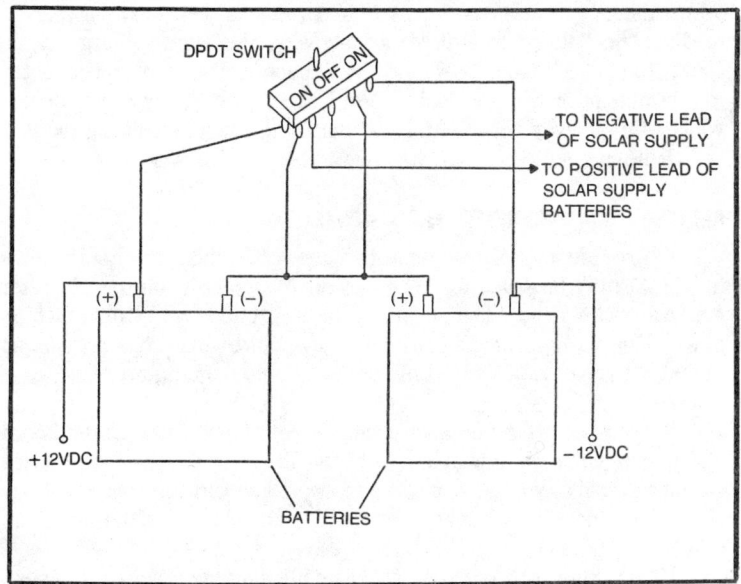

Fig. 4-28. Interconnection of rechargeable 6-volt batteries.

the polarity at the switch is not what you thought it was and is reversed. When you obtain a proper reading on the meter, the contact to which the red probe is attached is positive and should now be connected to the positive contact of one battery. The other contact lead is connected to the negative terminal of the same battery. Do the same for the other switch contacts when the lever is in the opposite position. Notice that the negative contact of one battery is attached to the positive contact of the other.

Now, place the solar cell bank in bright sunlight again and allow the batteries to start charging. Switch the supply to the other battery after about an hour and continue this process until a full charge is received. This may take many hours, depending upon the current ratings of the cells and the batteries. Connect the output of the batteries to a device which is within its power capabilities. If this load functions properly, your supply is operational. Any problems with operation of this circuit will be attributed to a reversed diode which should be evident during the intitial check-out procedure, a defective diode, defective solar cells, or defective batteries. The diode could open up after the system has been checked operational, so be suspicious of this component should the circuit suddenly cease to function. The diode should be rated to handle at least 1½ times the current which is available from the solar supply. The only other cause of circuit malfunction would be a broken wire or wires. This condition frequently occurs around the switch contacts but may also be found in the interconnections of the solar cells or the batteries.

When powering circuits from this supply, you are actually receiving power directly from the batteries. The 6-volt solar cell supply is merely feeding one battery at a time and allowing the stored energy to be connected in series and used as a 12-volt supply. The avid experimenter may want to go further and design some battery-powered switching device which will do away with the manual switching function, causing the two cells to be charged at present intervals by electronic means.

AMATEUR RADIO TRANSMITTER FOR 40 METERS

For the amateur radio operators in the readership, the circuit shown in Fig. 4-29 is provided for solar powered operations on the 40-meter amateur radio band. This is a crystal-controlled oscillator which is powered directly from the sun. Three miniature solar cells provide an output voltage of 1.35 VDC. A meter is provided so that power input may be calculated.

The circuit is built on a very small piece of perforated circuit board and mounted in an aluminum box, as shown in Fig. 4-30. The circuit board can be as small as you can make it, although L1 should not be mounted too close to any metallic objects such as the crystal or the side of the aluminum case.

L1 is wound by hand on a ¼ inch diameter slug-tuned coil form. These forms can be purchased from most hobby and electronic outlets; but make

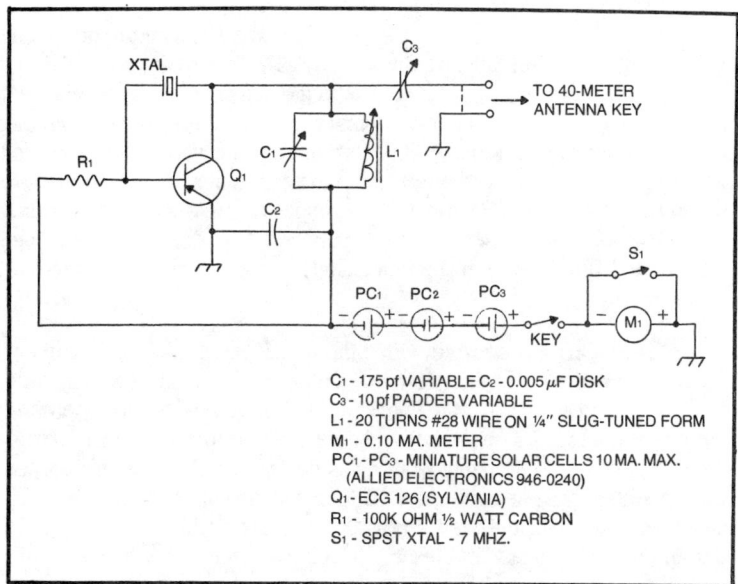

Fig. 4-29. 40-meter transmitter schematic.

certain the slug is ferrite and rated to at least 8 MHz. Some of these coil forms have iron slugs which will not work with this circuit.

Begin by mounting Q1 to the left-hand side of the circuit board, as shown in Fig. 4-30. The crystal may be directly wired into the circuit board, but a crystal holder is preferred. It will allow you to change transmitting frequencies without having to desolder the original component.

After Q1, R1, C2, and the crystal have been mounted, install L1 on the right-hand edge of the circuit board. This component is mounted vertically to allow for adjustment of the slug with a plastic tool. C1 and C3 are padder capacitors, although precision trimmer capacitors may also be used if desired. After all components have been wired and tested, each should be bonded to the circuit board using a drop of epoxy cement. If any of the components are mechanically unstable, the frequency of the transmitter will drift.

Still referring to Fig. 4-30, the wired circuit board is mounted on stand-off bolts in an aluminum box. The meter is mounted in this box along with the key jack and a coaxial cable connector. The jack and connector are chosen for the operator's individual needs. These will normally be a phone jack and a UHF or phono connector, but the builder may decide which components are best for his or her needs.

A small section of perforated circuit board is attached to the top of the aluminum case to which the solar cells are mounted using epoxy cement. Insulated leads are used for connecting the cells in the series circuit, with the positive lead connected to one side of the key jack and the other lead connected to the circuit board. Since the chassis forms the ground for this circuit, it is necessary to isolate the keying jack from the aluminum side it is mounted to. This can be done by inserting a grommet in the mounting hole, pushing the neck of the jack through this insulating material, and securing this assembly with the mating nut which holds the jack securely to the mounted grommet. S1 is provided to bypass the meter circuit, avoiding voltage drop. When a meter reading must be taken, the switch is opened and current flows through the movement, giving an indication of circuit current.

Check-Out Procedure. Once the circuit is wired and installed in the aluminum box, connect a 40-meter antenna and expose the solar cells to a light source. A 100-watt light bulb will do if you are not operating outside in the sun. Close the key and note the indication on M1. If current flows, the circuit is probably working as designed and only needs proper tuning. Release the key and close C1 to maximum compression. Open C3 about 75 percent of the way. Press the key again while monitoring the output on a radio receiver at least 100 feet away from the antenna. Adjust the slug in L1 for a maximum indication on the receiver "S" meter. Once the maximum reading has been reached, try adjusting C1 and C3 slightly for any improvement in output. Adjust L1 again, as the movement of these other controls may have an effect on its proper position.

If the circuit refuses to oscillate, try several different settings of L1, C1, and C3 until the note is heard in a shortwave receiver. Check to make certain the circuit has been wired properly and especially that the solar cells are connected in proper polarity. The meter is also a polarized device, so if it should tend to travel below scale, either the solar cells or the meter have been reversed. If the solar cells have been reversed, it is unlikely that the transistor will have been damaged; but if the circuit refuses to operate, you might try replacing this device.

It should be pointed out that this circuit requires an amateur radio license for operation. While it is an extremely short-range transmitter under normal circumstances, its output could travel several thousand miles given the right conditions. Only a person holding a bonafide amateur radio license may operate this circuit into an antenna.

For those who would like to go further, this circuit may be made to operate on other frequencies simply by changing C1, C3, L1, and the crystal. The Sylvania device specified is a VHF transistor and will easily cover the entire high-frequency amateur band using this same circuit design. C1 and L1 should be made to resonate at the desired operating frequency, while the crystal is chosen for the exact frequency desired. It may not be necessary to replace C3 with another capacitor, as its tuning range is adequate for most of the amateur radio bands.

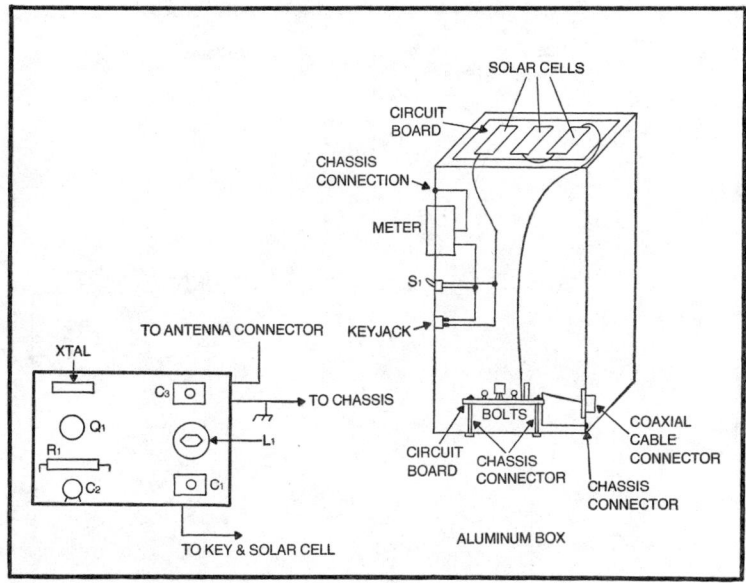

Fig. 4-30. Mounting of components on circuit board.

CODE PRACTICE OSCILLATOR

For those readers who are not yet amateur radio operators but who desire to be, it will be necessary to learn to send and receive Morse Code. To do this you will need a code practice oscillator; and in staying with the format of this book, the one shown in Fig. 4-31 is powered from the sun or a 100-watt light bulb.

This is a simple two-transistor oscillator which has a pleasing output note that is adjustable in range through variable resistor R1. Two different types of transistors are used. Q1 is an npn type, while Q2 is a pnp device that is connected directly to the 3.2-ohm miniature speaker.

No switch is provided because the code key performs this function. The circuit is not activated in any way until the key is closed. When this occurs, 1.35 volts from the three series solar cells is connected to the circuit which oscillates and feeds its output to the speaker. This is a directly coupled circuit. No transformer is used between the output of the transistors and the speaker.

Note the manner in which R1 is connected as indicated by the inset drawing. The right-hand contact is not used.

Figure 4-32 shows the component layout on a small section of circuit board. It may be possible to install everything on a 2-inch square section. Only five components are mounted to this board; while R1, the speaker, and the solar cells are mounted in, or to, a plastic enclosure. Insulated hookup wire is used to make the connection between the circuit board and these chassis-mounted components.

115

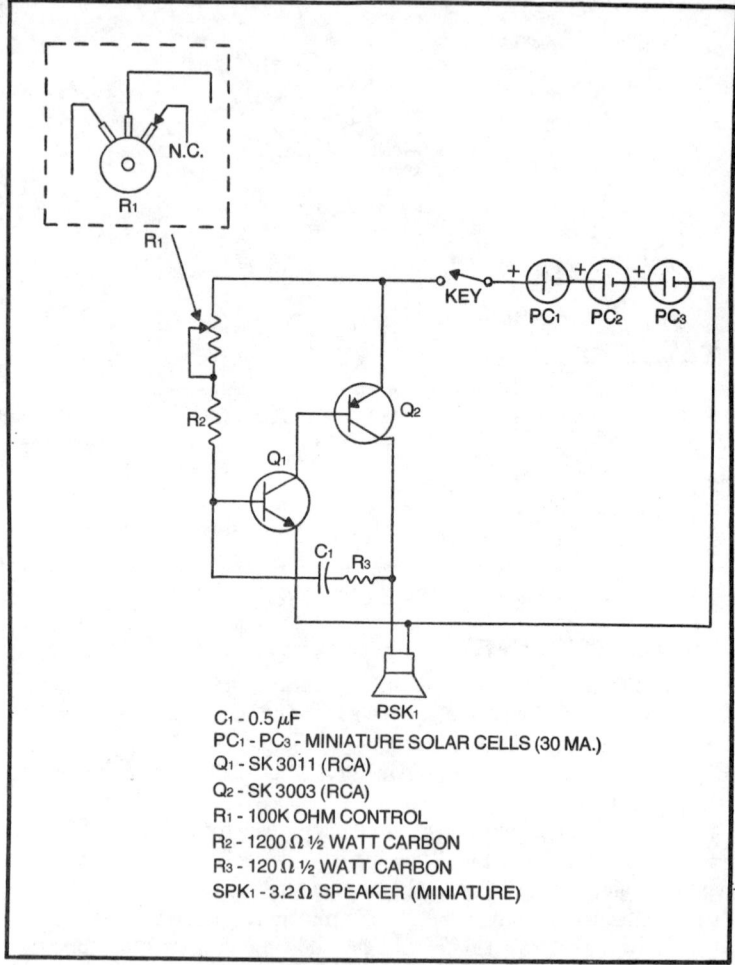

Fig. 4-31. Solar powered code practice oscillator circuit.

Figure 4-33 shows how the circuit board is mounted at the bottom of the plastic enclosure. R1 is installed through one side, while the speaker is mounted from underneath to the top enclosure cover. The three solar cells are epoxied to the top of the enclosure near the speaker. It will be necessary to drill holes or to make a cutout for the speaker. Many plastic enclosures have an aluminum front cover. This may be easily drilled to allow the output from the speaker to carry through. If your enclosure has an aluminum front cover, make certain the solar cells are insulated from it. A small section of perforated circuit board may be epoxied to the front of the aluminum and insulated hookup wiring run through holes to the circuit board and key jack.

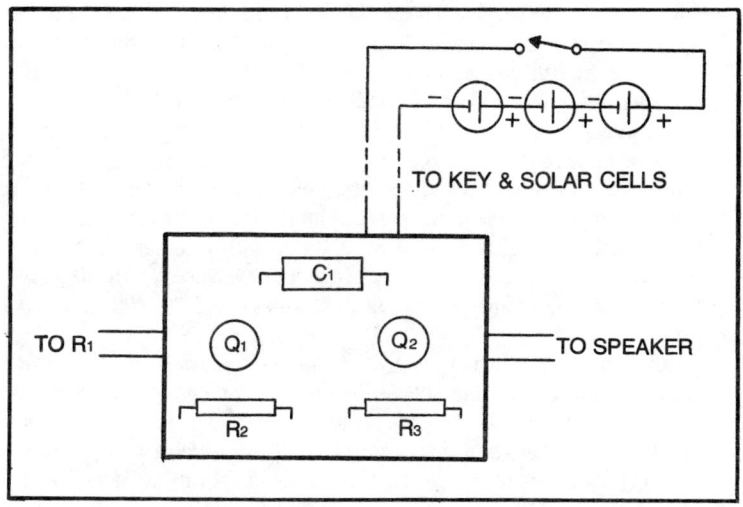

Fig. 4-32. Component layout for oscillator circuit.

Check-Out Procedure. Complete the wiring of the circuit board; then install the key jack, R1, the speaker, and solar cells in the enclosure. Do not install the circuit board in the same enclosure at this time. Make the various connections between the ciruit board and the chassis-mounted components using alligator-clip leads. If the circuit functions properly, the board may then be mounted permanently in the enclosure.

Expose the solar cells to a bright light source and close the key. You should immediately hear a tone in the speaker. Release the key. The tone

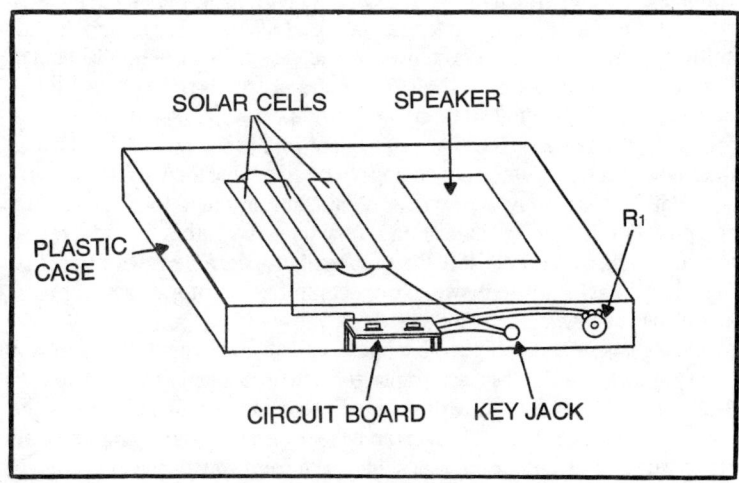

Fig. 4-33. Mounting of circuit board in plastic enclosure.

should cease immediately. There should be no decay of the signal. It should begin immediately upon closing the key and end when this device is open. Close the key again and adjust R1. This should raise or lower the frequency of the tone. Volume should be adequate for normal indoor use, although adding other solar cells in this string will provide for a substantial increase in output power.

If the circuit should fail to operate, open the code key and place the probes of a voltmeter across the positive and negative outputs of the solar cells. A total reading of about 1.35 volts should be obtained with the positive terminal being located at the connection to the key. If the negative terminal is at this point, reverse the solar cells and the circuit should operate.

Other problems which would cause the circuit not to operate properly, assuming that the solar cells are functioning and connected as shown, include a reversal of transistor leads, a defective speaker, or a bad solder connection between circuit elements. This is such a simple circuit, with only a few basic components, that it should operate properly on the first try.

To experiment further, try changing the value of C1 by paralleling it with other values of capacitance. You will be surprised at the variations of tones which may be obtained with this simple circuit. For those rainy days and nights, you may even want to install a 1.35 volt rechargeable battery in parallel with the solar cells. This battery will charge up during the daytime and will provide ample power even when adequate light is not available to obtain operational current directly from the solar cells.

PHOTOTRANSISTOR COMMERCIAL KILLER

Almost every book on photoelectric projects includes a circuit called a commercial killer which is designed to interrupt the sound on a television receiver whenever a commercial is aired. This interruption is not automatic and must be programmed by the viewer by directing a beam of light from a flashlight onto the surface of a photoelectric cell. All these devices are slight modifications of previous circuits which have been around for many years. They are equivalents of the day-night switch which is activated or deactivated depending on the amount of light available.

Rather than copy a previous circuit and attempt to make it look completely different by indicating a new use, the author has elected to admit that the commercial killer is simply a light-controlled switch just like the other light-controlled switch projects in this chapter. However, this switch uses a new component to activate the switching relay. Instead of a single transistor and a photocell or photoresistor, Q1, a phototransistor, is used to key Q2, a standard npn transistor, into a conductive state (see Fig. 4-34). This closes K1, a 9-volt relay whose contacts make and break the connection between the speaker and the television audio amplifier. In other words, this circuit offers a slightly different way of accomplishing a light-controlled switching function.

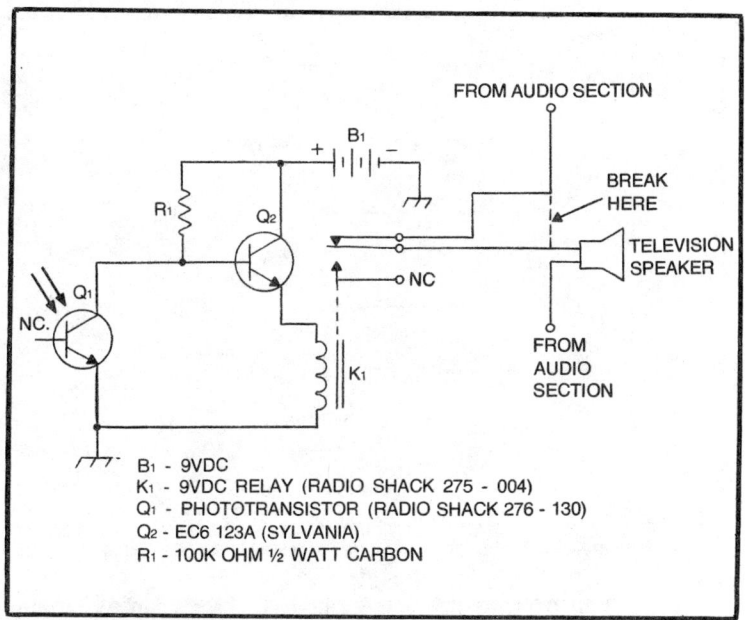

Fig. 4-34. Phototransistor commercial killer circuit.

Although a phototransistor has replaced the solar cell, or photoresistor, of earlier circuits and a resistor has been added, this project may be less expensive because the price of the phototransistor specified is less than $1 from Radio Shack. Most photoresistors and all common solar cells cost in excess of this amount.

Figure 4-35 shows how the components might be placed upon a small piece of circuit board. The 9-volt battery is also mounted to this board and will probably be the largest component found there. The relay switching contacts are connected to insulated leads which will run to the television speaker. This device is mounted in a small plastic or aluminum enclosure which is designed to be set atop the television receiver. Also shown in Fig. 4-35 is the mounting configuration of Q1. This device is located through a small hole drilled in the enclosure, allowing its photosensitive surface to be exposed for contact with the directed light beam. Q1 has three leads, but only two are used for connection to the circuit. The base lead may be clipped away.

Check-Out Procedure. Before the circuit board is permanently mounted in its protective enclosure, connect the battery and the phototransistor for the initial testing. This is a very simple procedure and actually requires no instruments, although an ohmmeter will be helpful. Connect this meter to the output lead from the relay. You should obtain a reading of zero ohms when the phototransistor window is covered. If the resistance indicated is infinite, then you have attached one of the leads to

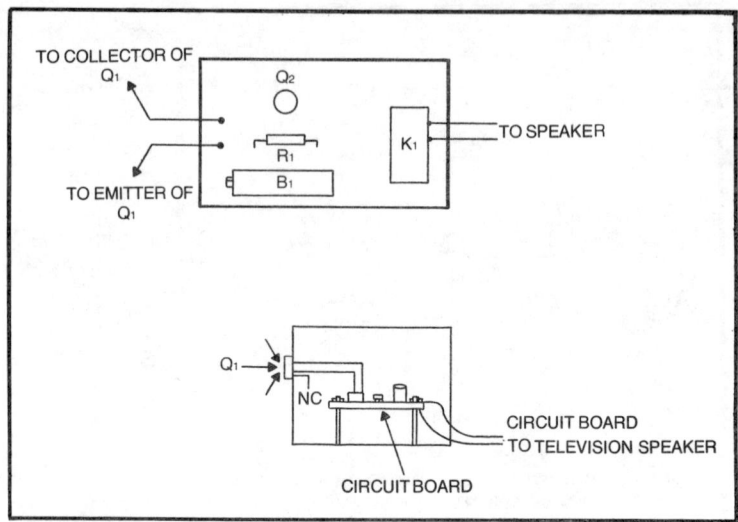

Fig. 4-35. Component mounting configuration on perforated circuit board.

the normally open contact instead of the normally closed element. Reverse this lead and check again. When you have a reading of zero ohms, the leads are properly attached.

Now, remove the cover from the phototransistor and expose the treated surface to the light source. You should hear the relay click and the ohmmeter reading should indicate infinite resistance. If this does not occur, the relay is not keying. This could be due to a dead battery. Check this with a voltmeter and replace it if the reading is questionable. The only other conditions which could bring about inoperation would be an improperly installed transistor, defective relay, or improperly connected battery. If the positive and negative battery leads have been reversed, one or both transistors could be damaged. Q2 is most likely to be damaged by this reversal and should be checked and replaced if necessary.

The connection to the television receiver speaker is quite simple. Locate one of the leads which is soldered to a speaker terminal. Clip it and attach the relay leads to the two sections. Solder these connections and wrap them with insulating tape. When the circuit is not activated by the light beam, sound will be heard from the speaker. When light strikes the treated surface of the phototransistor, the relay will engage and break the connection between the audio amplifier and the speaker. It will probably be necessary to mount a small section of cardboard tubing around the phototransistor, as shown in Fig. 4-36, to shield this device from normal room light. The beam from a powerful flashlight may be aimed directly down the mouth of the tube in order to activate the circuit.

The current rating of the contacts in the relay specified in the schematic is only 1 ampere at 115 volts ac. Many televisions draw more operating current than this; so if this switching circuit is to be used to control the on-off power function of the receiver instead of the speaker sound as outlined in this project, it will be necessary to use a relay with higher contact ratings or to drive such a relay, as was done in a former project, by controlling its power supply through the contacts of K1.

AUDIBLE LIGHT METER

The simple light meter circuits described earlier in this chapter all used meters to provide indication of relative light levels. The circuit shown in Fig. 4-37 registers light intensity through audible clicks rather than on a meter face. For most applications, this circuit is more experimental than practical; but it is an interesting exercise in converting from light energy to electrical energy to audible energy. The circuit can also be used as a light-controlled metronome and counting device.

The output from this electronic circuit is heard as clicks. These sounds are evenly spaced and will increase as light intensity increases. The photoresistor, PC1, is the controlling element here. As more light strikes its surface, the rate of clicks increases. R2, a variable resistor, is used to establish the general range of frequency. This may be eliminated or even placed in series with the photoresistor for variation on this basic circuit.

A 9-volt battery supplies operating power for this circuit but may be replaced with any of the 9-volt AC power supplies designed to replace batteries of this type in transistor radio and other types of electronic circuits. The battery is mounted to the inside of a plastic or aluminum case, as is shown in Fig. 4-38.

Figure 4-38 also shows the wiring of the circuit board. The two largest components will be the output transformer, T1, and a 120 μF capacitor, C1. Make certain that the capacitor is installed with its positive lead connected to the base of Q1. Its negative lead is connected directly to

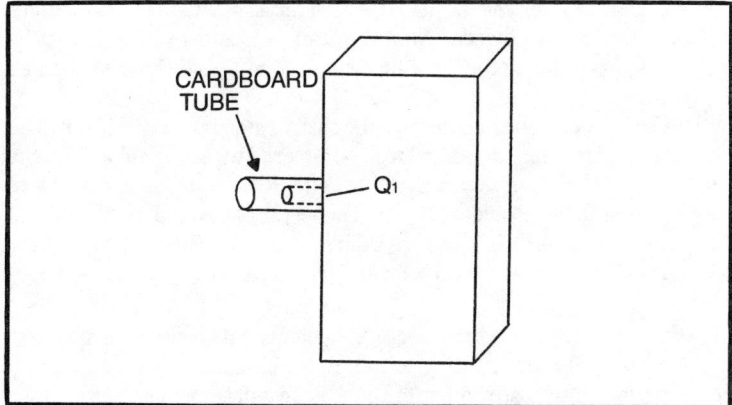

Fig. 4-36. Mounting cardboard tubing around phototransistor as light shield.

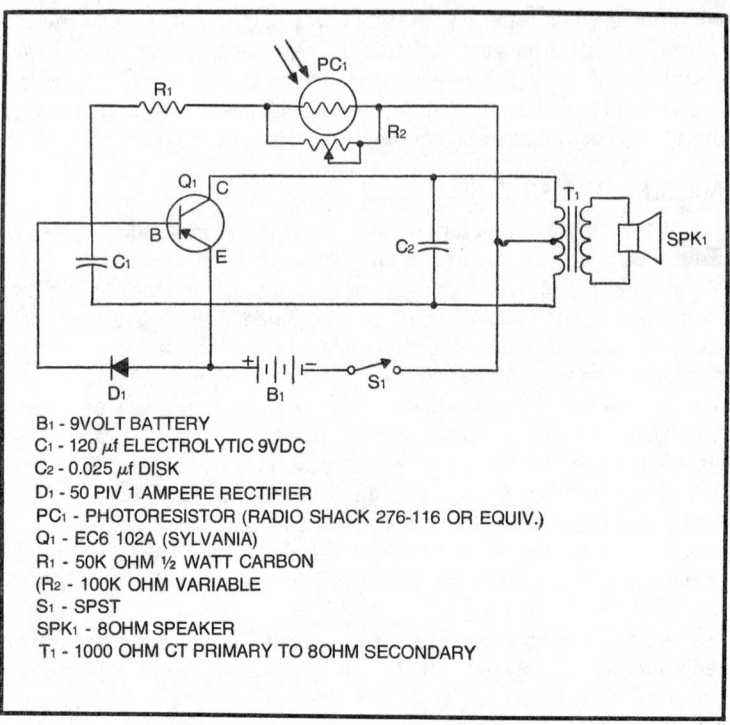

Fig. 4-37. Audible light meter circuit.

one side of the primary winding of T1. It is very easy to determine the primary and secondary windings of this latter device. The primary winding is center-tapped. It contains three leads. The secondary has only two.

Wire the circuit board as shown with C1 mounted horizontally on the left side and T1 on the right. The smaller components are mounted between these two devices. Note the leads which consist of four pairs emanating from the circuit board for connection to the chassis-mounted components.

Install the components through the chassis as shown in Fig. 4-38. The photoresistor mounts on top and may be covered with a cardboard tube to partially shield it from normal lighting conditions. This same procedure was followed on the previous project. The length of the cardboard tube will determine the amount of effective shielding and the accuracy with which the control lighting source will have to be directed.

Check-Out Procedure. Once the components have been mounted in the enclosure, make the connections to the circuit board as shown, but do not permanently fasten down the perf board before testing. During this testing procedure, make certain the component leads on the bottom side of the board do not become shorted if an aluminum chassis is used as an enclosure. Connect the battery and turn S1 to the on position. Shield PC1 from any light. Rotate R2 until a slow series of clicks are heard in the output speaker. Now, expose the photo resistor to a source of light. The clicks should immediately increase in rate, or frequency. By further adjusting R2 and controling the intensity of the light source, the click frequency can further be adjusted.

If the circuit refuses to oscillate, suspect first the polarized devices such as B1, D1, C1, and Q1. If any one of these devices is reversed, the circuit will not function properly or at all. The battery, of course, must be a fresh one. Be sure to check this, as many frustrated hobbyists have become even more frustrated when learning that the project which took many hours to build and many days of troubleshooting when it didn't work properly did so because of a bad battery. Even though the battery has been recently purchased, it may still be defective. Check any battery purchases with a voltmeter or one of the inexpensive commercial battery checkers available for a few dollars.

The light source can be varied in intensity by slowly increasing the voltage to a standard flashlight bulb mounted in the mouth of the cardboard tube; or the beam of a flashlight may be used, holding the reflector closer and closer to the tube opening to increase the click frequency. This device does not draw a lot of current, but many experimenters may wish to power it with a 9-volt battery replacement power supply which can be purchased for less than $10.00 from many hobby outlets. Other voltages may also be used to operate this circuit, although the maximum level should not exceed 12 to 15 VDC.

LIGHT-POWERED/LIGHT-CONTROLLED PULSE OSCILLATOR

An earlier project in this chapter featured an electronic organ which was controlled by light intensity through a photoresistor. The circuit

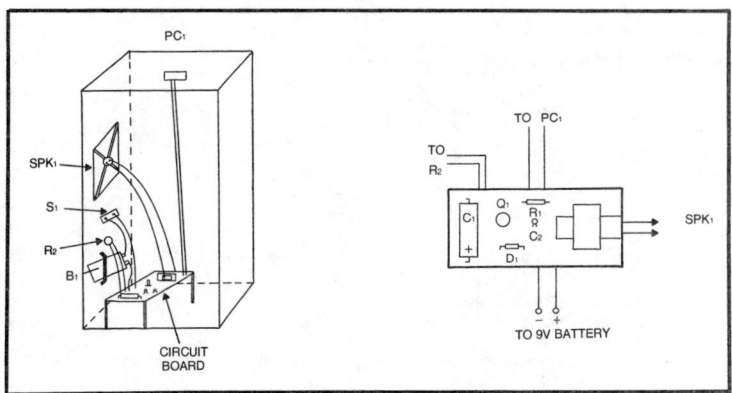

Fig. 4-38. Component placement on circuit board and enclosure layout.

shown in Fig. 4-39 is an electronic pulse generator or tone oscillator which can be used for many purposes that require a variable audio frequency output. This project can also be used as a tone organ which is powered directly from light energy and is controlled in frequency by light intensity. The output is heard from a miniature 8-ohm speaker connected to the circuit through a miniature output transformer. The tone is varied by directing a light source over the photoresistor or moving your hand back and forth over this device, creating a shadow effect.

Figure 4-40 shows the component layout on a small section of perforated circuit board. The largest item is T1, which is mounted on the left-hand side. The remaining four components are mounted near the right-hand side. Three sets of leads are brought off this board; one pair to the speaker, another to PC7, and the third pair to the solar cells and S1. These components are mounted in a small plastic or aluminum case similar to the type used for previous projects.

Install the components as shown on the circuit board layout. Make certain that the leads are plenty long in order to reach the components which are mounted in the enclosure. Any excess lengths can be easily clipped away; but leads which are too short will often have to be replaced entirely, starting at the circuit board connections.

Figure 4-41 shows the all-important component placement in the plastic or aluminum enclosure. Plastic is preferred for this particular project because the solar cells are mounted on the back. If aluminum is used, an insulating material must be epoxied to the back section to avoid shorting the connections. Make ample holes for the speaker. Alternately,

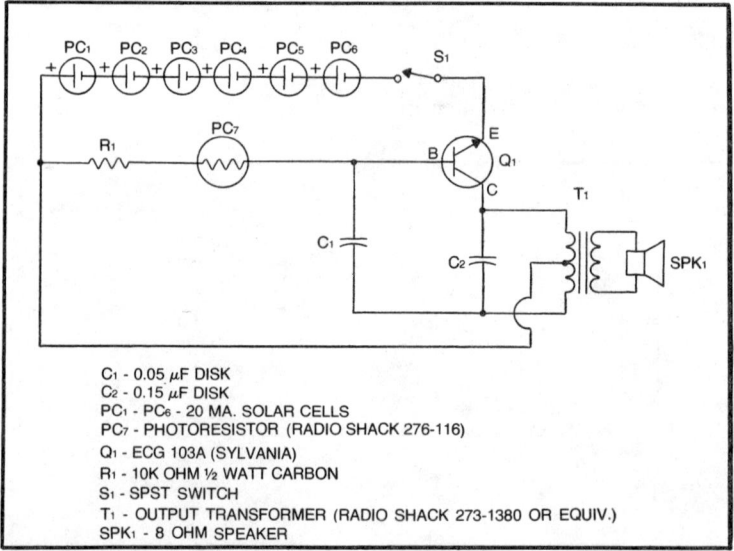

C_1 - 0.05 μF DISK
C_2 - 0.15 μF DISK
PC_1 - PC_6 - 20 MA. SOLAR CELLS
PC_7 - PHOTORESISTOR (RADIO SHACK 276-116)
Q_1 - ECG 103A (SYLVANIA)
R_1 - 10K OHM ½ WATT CARBON
S_1 - SPST SWITCH
T_1 - OUTPUT TRANSFORMER (RADIO SHACK 273-1380 OR EQUIV.)
SPK_1 - 8 OHM SPEAKER

Fig. 4-39. Pulse oscillator circuit.

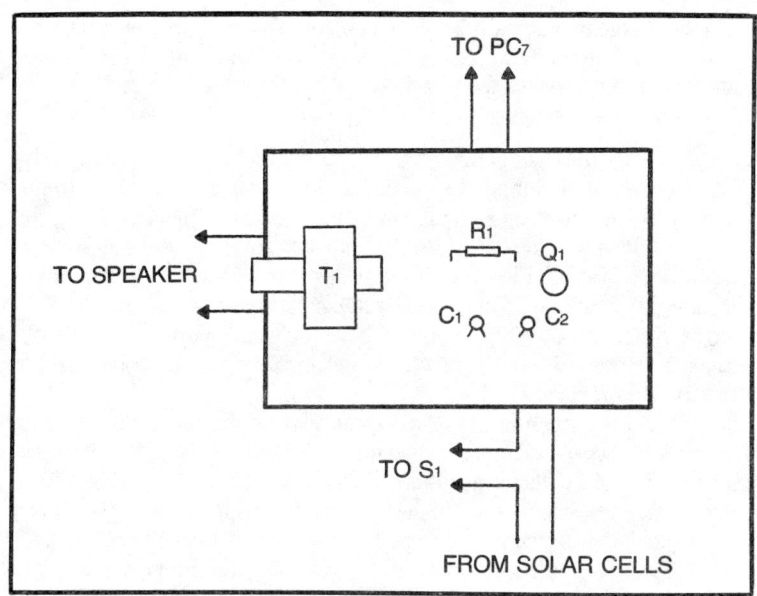

Fig. 4-40. Component layout of pulse oscillator.

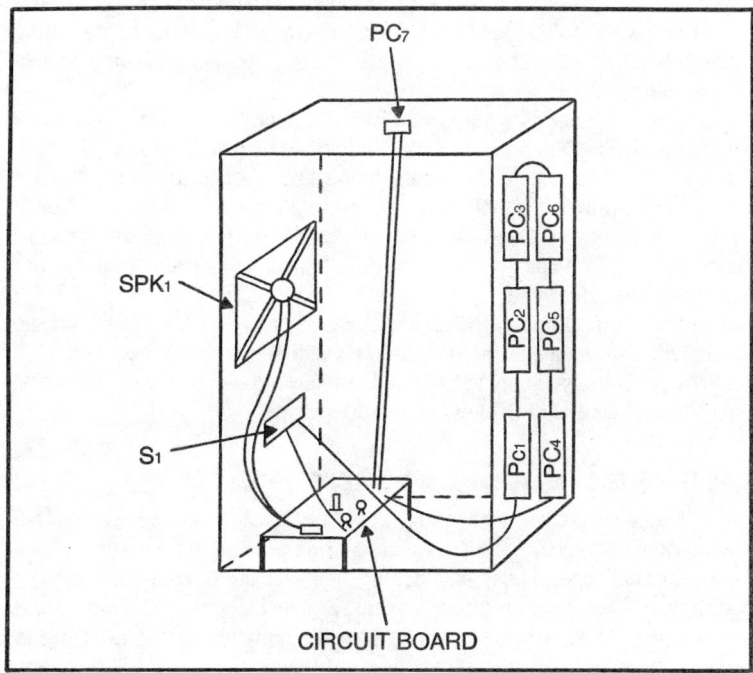

Fig. 4-41. Placement of components in plastic or aluminum enclosure.

a circular hole may be cut, the speaker installed, and a screened protector placed over the paper diaphragm. S1 is mounted just beneath the speaker and is attached between the solar cell supply and the circuit board.

Check-Out Procedure. As with many of our previous projects, the pulse oscillator circuit board is installed with small bolts and nuts to the bottom of the plastic or aluminum case. If plastic, the board may be mounted flush with the enclosure. If aluminum, it must be separated by the bolts, as shown in Fig. 4-41. It is best, however, not to permanently anchor the circuit board until the check-out procedure is completed. This saves you the trouble of having to laboriously remove the board from its mounts to recheck the wiring should the circuit prove inoperational the first time power is applied.

With the switch in the off position, expose the six solar cells to a bright light source. The photoresistor may also be exposed to the same source. Turn S1 to the on position and you should immediately hear a clear tone. Place your hand between the light source and the photoresistor, and the frequency will vary in direct proportion to the intensity of the light which is allowed to pass onto the treated surface. The frequency will also vary slightly in proportion to the light which strikes the solar cells.

If the circuit fails to work, recheck your connections of Q1 and the solar cells. The only other source of problems would be in broken hookup wire or a defective speaker. Once the circuit is known to be operating properly, anchor the board to the bottom of the enclosure and your project is complete.

One of the uses for this circuit is as a conversation piece. Placed on a coffee table inside a home or on a picnic table or bench outside, the tone will constantly change in proportion to the amount of light present. Inside, artificial lighting will supply enough power for an output which varies each time an interested onlooker moves close to examine the unusual contraption. This usually results in their blocking a portion of the light to the photoresistor, causing the tone to change. As they walk away, it changes again. Outside, the same sequence of events can occur, but the tone will also change as the sun rises and sets or goes behind a cloud. With a little practice, you may even be able to play simple melodies by moving your hand back and forth over the photoresistor.

SOLAR-POWERED WATCH FOR LESS THAN $2

Today, watches which are powered by the sun are attractive commodities but may cost two-, three-, or even four times as much as a conventional type. These watches are not actually powered directly by sunlight. They have an internal rechargeable battery which stores the power generated by tiny solar cells from sunlight. These watches are just like other electronic timepieces; they get their power from a battery and the battery gets its power from the sun.

The title of this project may be a bit misleading. The $2 price for a solar-charged watch assumes that you already own an electronic digital watch and that you have been able to locate a rechargeable battery. Standard watch batteries are not rechargeable, so you will have to check with the mail order electronic houses for one which is. Now, assuming that you have a digital watch and rechargeable battery which will fit it, you can *convert* to solar charging for less than two dollars.

First of all, purchase the cheapest solar cell you can buy. Its current rating is totally unimportant. You may even be able to pick up a few broken pieces of solar cell at a hobby store for a few cents. A broken cell will work fine, because in order to make this conversion you must break a solar cell into many tiny pieces. Figure 4-42 shows the schematic diagram of the circuit, which is merely three solar cells in series producing a voltage of 1.35VDC. These three cells are made from a single solar cell device and should measure no more than 1/16 inch in length and diameter. Figure 4-43A shows a small section of solar cell. This is the side which receives the light rays. Most cells are a parallel combination of many individual cells which have been combined to produce a single unit with a useable amount of current output. The lines shown in Fig. 4-43A are actually small strips of metal foil which connect that cell portion to a central strip used as the negative output terminal. Figure 4-43B shows how a single cell segment is isolated from the previous section shown. This may be done by snipping away small bits of the unwanted portion with a pair of diagonal cutters. This is tedious work and much breakage will occur, but you need only a tiny section to work with. To complicate things, you need three sections, all of approximately the same size. The output from each section will be the same as for the entire cell, or about 0.45 VDC. The current output is a different matter altogether. A single cell section only puts out a few microamperes of current, but this is all that will be required to charge

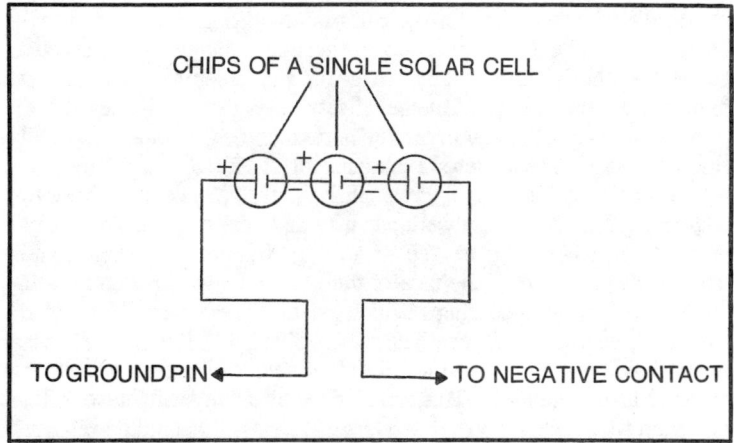

Fig. 4-42. Solar watch conversion circuit.

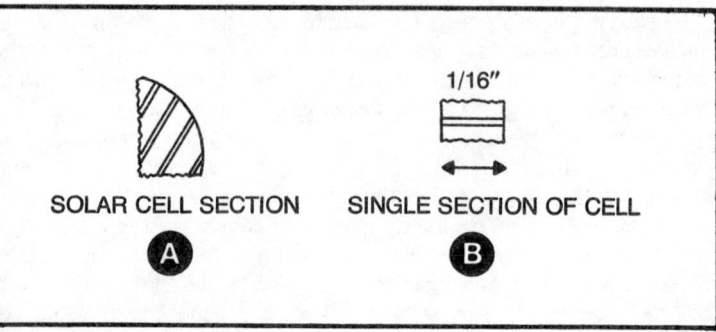

Fig. 4-43. Solar cell sections.

the watch battery. Figure 4-44 shows how these cells are combined in series. This is extremely detailed work and should not be attempted by the neophyte builder. A magnifying viewer will be of great help in making these delicate interconnections.

Purchase a length of aluminum-foil tape from a local hobby store. (This is often used for burglar alarm purposes and is applied to windows. It has an adhesive backing which allows it to stick to the glass. Should an intruder break a window, the foil tears and opens a circuit, causing an alarm to sound.) This material is normally sold in rolls of 25 feet or more. For this purpose, only an inch or so will be needed.

Cut thin sections of this foil with a pair of small surgical scissors. The diameter of this tape should be about the same as the diameter of the foil on the cell segments. Put a dab of solder on the segment strip by holding a tiny soldering iron to the surface and allowing the solder to melt and flow. Once this is done, the alarm-tape strip is pressed on top of the solder, heated by the iron, and bonded to the segment. Move onto the next cell and apply solder to the bottom side, as shown in Fig. 4-44. The strip which has been attached to the top of the former cell has its other end connected to the bottom of this one. Now, move to the other end of this second cell section, connect another sliver of foil tape to the top and solder its other end to the bottom of the third section. Another strip of foil is attached to the top of the third segment and yet another to the bottom of the first one. Congratulations, you have just completed a sub-miniature 1.35VDC solar charger.

There is no easy way to accomplish the task of soldering these tiny segments. It is frustrating work and may take many hours. The finished product will have to be treated with kid gloves until the string can be permanently bonded to the face of the electronic watch. Figure 4-45A shows the proper mounting position on the watch face. Most of the watches have a plastic cover which is a good insulator. It will be necessary to punch two tiny holes on either side of the solar cell mounting location in order to insert the leads. This can be done by using a miniature drill bit designed to bore holes in printed circuit boards. It should be relatively easy to obtain one of these from a hobby store.

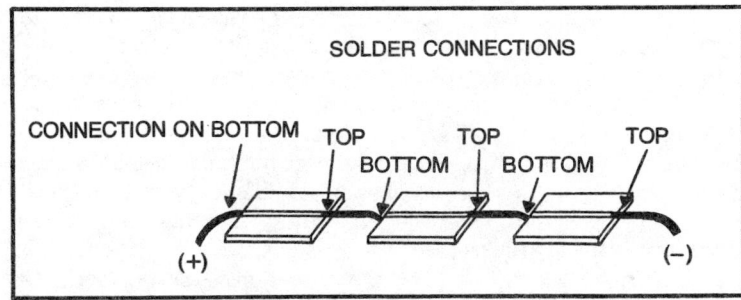

Fig. 4-44. Series wiring of three solar cell segments.

Figure 4-45B shows the back of the watch. The internal circuitry of most common electronic timepieces will be very similar to this drawing. The entire electronic circuit will drop out of the case when the back is removed. This should be done before drilling the front cover.

Referring to Fig. 4-45B, the crystal is normally contained in a hollow well at the top of the circuit board. You must drill two holes in this well that correspond with the holes in the front face. It is often better to drill these holes at one time from the back side. Now, obtain two lengths of the finest enamel-coated copper wire you can find. This can often be obtained by unwinding the turns of a small rf choke. Scrape the insulation for 1/30th of an inch from one end of each wire. Insert each wire through one of the holes in the well and through the front cover. You must now wrap the wire ends with the two foil leads from the solar cell supply. These are soldered and excess material is snipped away. Again, a lighted magnifying viewer will be of great help.

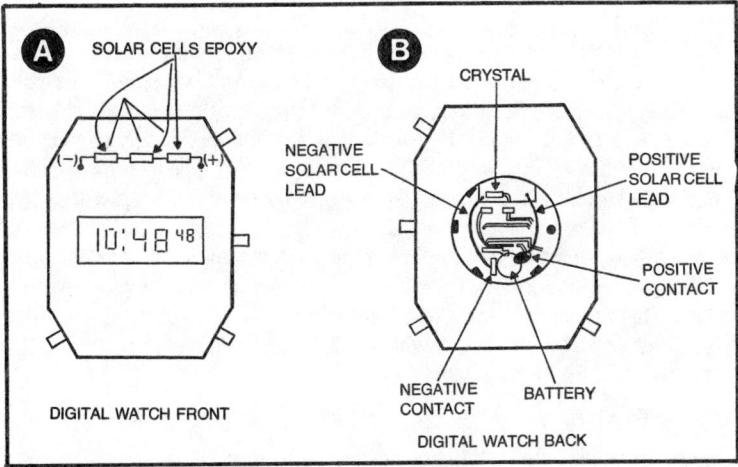

Fig. 4-45. Front and back component placement and wiring of electronic watch.

Now, route these two wires to their proper points in the circuit. The positive contact will most often be attached to a small latch which holds the battery in place. This is identified as the "positive contact" in the drawing of 4-45B. The negative connection is usually located on the opposite side of the battery well and is so labeled in the drawing. Scrape the insulation from the two wires for a small distance from the remaining ends and solder them to these contact points. If your watch is different, a voltmeter will help you identify the positive and negative battery contacts.

Remove the battery from the watch and expose the solar cells to a source of bright light. Now, connect the positive and negative probes of a voltmeter to the positive and negative contacts of the watch surface. You should obtain a reading of a little over 1 volt. If not, you have a broken contact, either between the solar cells or between the solar cell string and the watch contacts.

Check-Out Procedure. Once you have established that the solar cell supply is providing charging current to the battery contacts within the watch, re-insert the battery, close the latch over its top, and snap the back lid to the case. Set up the time and date functions on your watch according to the manufacturer's directions. It should be working in a normal manner. If not, re-examine all connections that have been made. Even if the solar cell charger is not functioning, the watch should work, assuming that the battery being used has a reasonable charge. With a voltmeter, again measure the output from the three solar cells by attaching the meter probe to the soldered contacts on the front of the case. When exposed to bright light you should read over 1 volt. Now, cover the sides of the solar cells with a thin coating of bonding material such as the "super glues" found at discount stores. This is better than expoxy because it often dries clear. Cover the exposed solder contacts with this material as well. Your project is complete.

This project has one major flaw; it may take you several years to know for sure whether your solar cell charger is actually working. This will depend upon the storage capacity of the battery and the current requirement of the watch. If your watch never stops running, you can assume that the solar supply is fully operational. For those individuals who can't wait, remove the watch battery and connect it to a small light-emitting diode or a miniature panel lamp, allowing it to discharge most of its energy. Re-insert it into the watch, making certain that it does not have enough stored energy left to operate the timing circuit. Now, allow the solar cells to recharge the battery by leaving your watch face exposed to a bright light source for about eight hours. If the watch begins to operate properly, your project has been a success.

Caution: Make certain the three solar cell segments are putting out the correct voltage before applying the bonding glue. This material effectively makes the cells a permanent part of the case, and removing them for repairs will be next to impossible.

There you have it—a solar charged watch for less than $2, the price of a single cell. The author fervently hopes that no readers go completely crazy from the minute work and almost surgical skill that is required to connect the cell segments and internal watch wiring.

150 VOLTS DC FROM A 1.35-VOLT SOLAR CELL SUPPLY

The circuit shown in Fig. 4-46 is intended for experienced builders only. It is not highly complex but requires certain building skills, such as winding a transformer, that only come through experience and training. The circuit is known as a DC-to-DC converter and takes the output from a solar power supply and rechargeable battery, both rated at 1.35 VDC, and converts it to an output of about 150 VDC, from which you can safely draw about 10 milliamperes of current. This assumes that the rechargeable battery has a rating of about 2 amperes. The solar cell supply will directly power this circuit up to a 2-ampere input when they are subjected to bright sunlight. For these applications, the battery may be eliminated, but the current drain from the 150-volt output should not exceed 7 milliamperes.

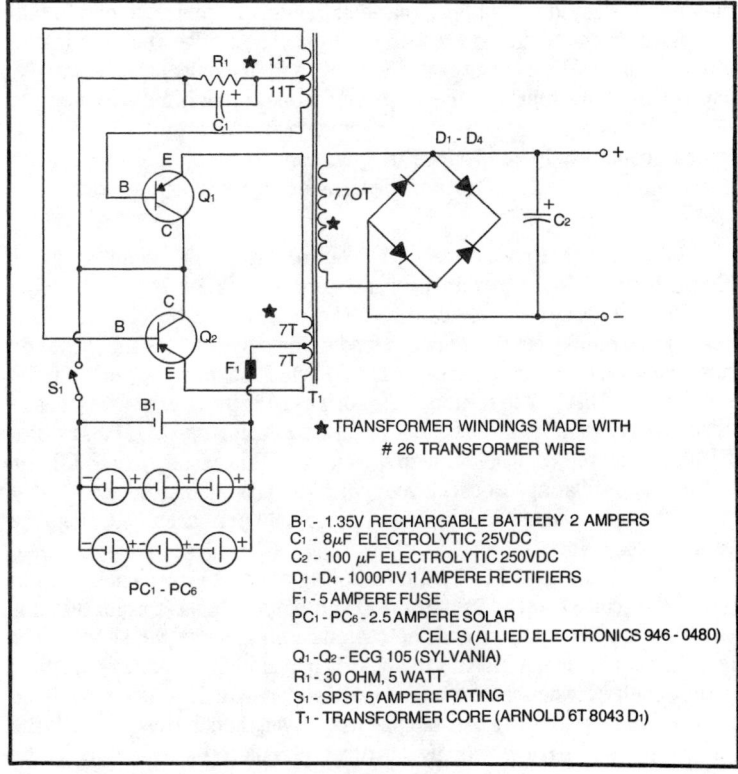

Fig. 4-46. Schematic diagram of 150-volt supply.

This will represent a current drain from the solar supply of about 2 amperes.

Without going into great detail about circuit principles, the DC current is switched by the two transistors between positive and negative at a very high frequency. This causes the DC current to appear as alternating current to the transformer. Only alternating current can be used with standard transformers. The 1.35-volt value of the battery and solar cells is stepped up in the secondary winding to a value of about 150 volts, where it is rectified and filtered. The output is pure DC.

The transistors must be mounted on heat sinks which will effectively allow them to get rid of heat which has built up in the switching process. The most difficult portion of this project is in winding the various turns around the circular core of this special transformer material. One winding consists of 22 turns, center-tapped. Another is wound with 14 turns of wire and also center-tapped. The secondary winding consists of 770 turns of wire. All of these windings are located on one toroidal core. All of the windings are made with #26 transformer wire.

This project is submitted as a starting point for experimenters who may be interested in building solar cell power systems designed to take the place of conventional electric utilities. The bridge rectifier in this circuit can be done away with and the output could drive some AC appliances. Unfortunately, the frequency of this output is much higher than 60 Hz and is not suited to most common appliances in the home. The voltage could easily be dropped to a more suitable value of 115 VAC by decreasing the number of turns in the transformer secondary winding.

Check-Out Procedure. The entire circuit is mounted in an aluminum case. Make certain the transistors are isolated from the metallic heat sinks unless these devices are also isolated from the chassis which serves as circuit ground. Connect a load to the output of the supply but do not turn it on until you have activated S1 which supplies power from the solar cells and battery to the input. As soon as S1 is thrown, you may hear a shrill whine, which is the frequency of the switching rate of Q1 and Q2. If F1 blows, there is a problem with the circuit. This usually means Q1 and Q2 are not oscillating. Recheck your wiring and try again. Replace F1 only with a 5-ampere fuse. One of higher rating might permanently damage the transistors.

If the supply seems to function properly when S1 is thrown, switch on the load circuit powered from the 150 volt output. Make certain that this latter circuit does not exceed the current limitations of the supply. The load circuit should function normally. You may notice a change in the pitch of the oscillator when the circuit is placed under load. This is normal, but if the oscillation grinds to a halt and the fuse blows, then the load exceeds the current limitations of the supply and must be replaced with another, more suitable circuit.

SOLAR-POWERED 2-METER
CONVERTER FOR THE AMATEUR RADIO OPERATOR

This is an ideal project to be used with a solar powered AM radio. It will allow you to listen in on a portion of the 2-meter amateur band through your solar powered or standard AM receiver. Shown in Fig. 4-47, the circuit consists of a 48-Mhz oscillator which mixes its output with the signal of another oscillator whose output is in the broadcast band. The 48-Mhz oscillator has an output of three times the crystal frequency, or about 144 Mhz. This output is connected to the bottom of a loopstick antenna which broadcasts to a nearby AM radio.

L1 and the antenna are designed to receive 2-meter signals from a station or repeater near you. This circuit provides good sensitivity and you may be surprised at the distant stations you bring in. Six solar cells provide power for the entire circuit. Alternately, these may be used to charge two 1.35-volt penlite cells wired in series. This would take the circuit into a more practical realm.

Figure 4-48 shows a component layout diagram. The entire circuit, with the exception of the switch and solar cells, is located on a single piece of circuit board and is wired as indicated. Attempt to keep L1 and L2 separated by the entire length or width of the board so that these devices do not interact. All wiring leads should be kept as short as possible and any components which tend to be mechanically unstable should be bonded to the board with a drop or two of epoxy cement.

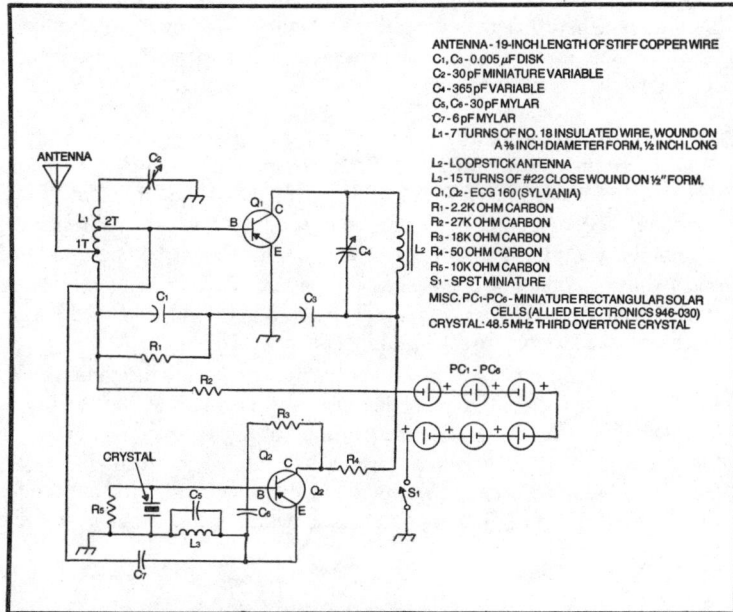

Fig. 4-47. Amateur radio converter circuit.

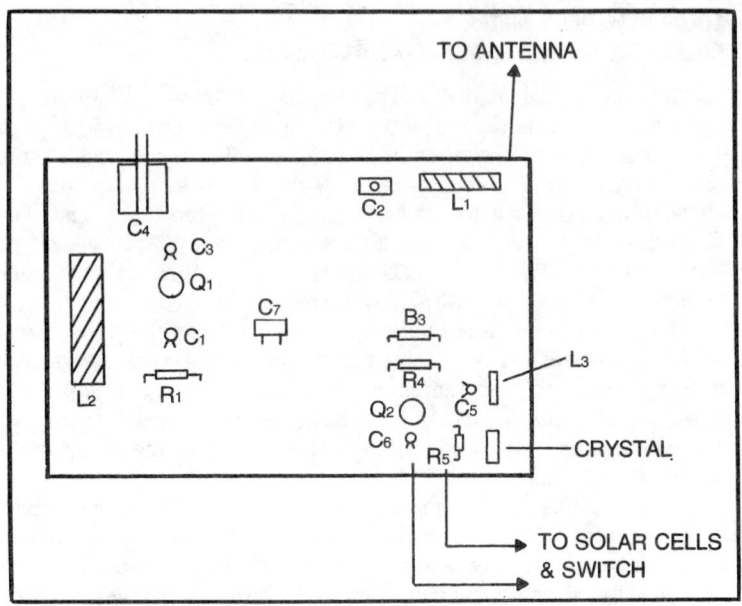

Fig. 4-48. Converter layout on perforated circuit board.

This circuit is best tested outside its aluminum enclosure until it has been established that it is operational. Figure 4-49 shows this cabinet, which houses the circuit board and provides a mounting area for the solar cells, switch, and antenna.

Check-Out Procedure. With S1 in the off position, locate the solar cells so that they absorb the energy from a bright source of light. Turn on a transistor radio and place it near the circuit. Now switch S1 to the on position. The crystal should tune the frequency of a local repeater. Alterately, you may have a friend who is an amateur radio operator send out a transmission while you align the circuit. Adjust C2 for maximum signal strength of the received signal as heard in the radio. C4 will allow you to adjust the output to a desired spot on the AM dial.

If the circuit does not operate properly, try placing the radio nearer L2. Also, check all internal connections, especially around the oscillator. If you have a receiver which will cover the 144 Mhz range, you may listen for the output of the oscillator, which will be just above or below the frequency being received. This will depend on the crystal you choose for the oscillator. The following formula shows how the frequency of the crystal is determined:

$$\text{Crystal Frequency} = \frac{F_1 - F_2}{3}$$

F1 is the desired 2-meter signal to be received, and F2 is the desired place of reception on the AM dial.

EXAMPLE:

$$F1 = 147.00 \text{ MHz.}$$
$$F2 = 1.00 \text{ MHz. } (1000 \text{ kHz.})$$
$$\frac{147.00 - 1.00}{3} = \frac{146.00}{3} = 45.333 \text{ MHz.}$$

HIGH-CURRENT SCR SWITCH

Most of the projects used for switching on and off other electronic circuits have been limited to relatively low-powered loads unless an additional relay was added with contacts rated for the greater current demands. Relays are sometimes bulky and expensive; and being mechanical devices, they are more subject to failure than would be a solid-state equivalent. Figure 4-50 shows a circuit which will easily handle a 1000 watt DC load. This could be 10 100-watt bulbs, two 500-watt bulbs, etc.

The only moving part in this circuit is R1, which is a variable resistor. All of the switching functions are handled by a silicon-controlled rectifier, SCR1, which is activated by a single photoresistor. When no light is present at PC1, the SCR is not in a conducting state and no current

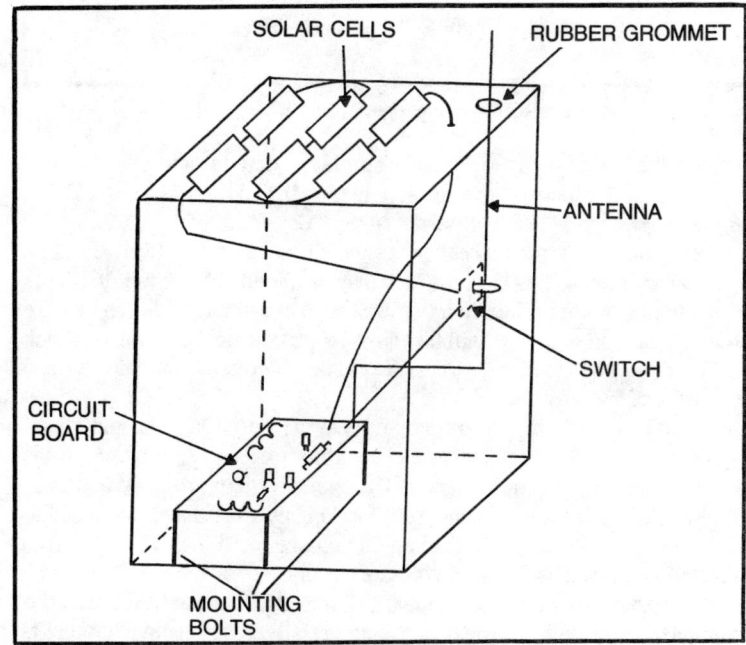

Fig. 4-49. Mounting components in aluminum enclosure.

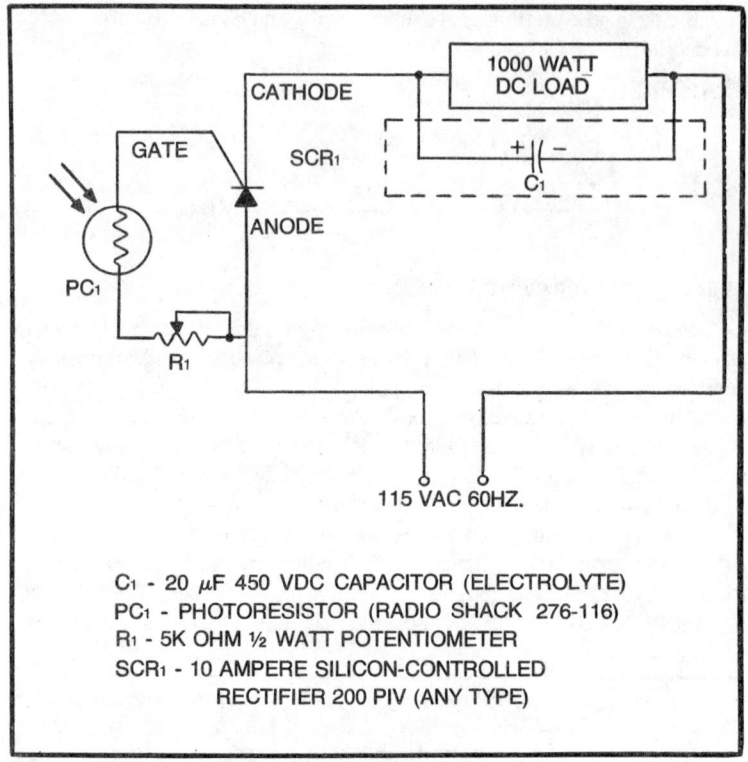

Fig. 4-50. High-current SCR switch circuit.

reaches the load. When the photoresistor is subjected to a focused beam of light, the SCR conducts, the AC supply is rectified, and 115 volts DC is applied to the load. This will be pulsating DC, and a slight flicker may be noticed from any incandescent lights which make up the load. C1 is an optional addition to this circuit which may be added to filter the AC ripple component from the DC output of the SCR. This output will then be pure DC, but the voltage value will be equivalent to the peak AC value, which will be about 150 volts. If the capacitor is added, make certain the DC load is capable of operating at 150 VDC.

SCR1 is a 10-ampere device and may be purchased through many outlets. Industrial grade devices can be rather expensive, although this is not an extremely high rating for an SCR. Some devices are rated to handle hundreds of amperes of current. Your local hobby store may have this type of device in ten-, fifteen-, or even 20-ampere versions. The industrial surplus catalogues list these devices for a couple of dollars.

Due to the amount of current which is passed through the SCR, the component body begins to heat up. Being a relatively small device, there is not enough surface area to its case to adequately get rid of the heat to the surrounding air. A *heat sink* must be used as a mount for the SCR. This is shown in Fig. 4-51. A heat sink of this type is simply an aluminum plate with fins which do a very good job of dissipating heat into the air. The SCR is mounted through a small hole in the center of the heat sink and is usually electrically isolated from the metal conductor of the heat sink by means of mica washers. This material provides good electrical isolation but will continue to conduct the heat from the SCR body to the metal fins. Standard mounting kits usually include all of the hardware for mating the SCR with a common heat sink. This kit will also include a silicone paste which is smeared over the device, the heat sink, and the washers. Do not neglect to use this paste, as it assures a good thermal bond between all components.

After the SCR is mounted to the heat sink, you may begin the wiring procedure. Since nearly 10 amperes of current will be conducted through this circuit, use standard electrical wiring of #14 gauge or larger to avoid conductor heating. Referring to Fig. 4-52, the threaded shaft of the SCR is the anode connection, while the cathode is the longer of two contact shafts at the other end. The shorter contact is the gate. Normally, this entire assembly will be mounted atop a small chassis which may be covered by a screened enclosure to allow for adequate circulation of air to the heat sink. The mounting configuration will be left up to the reader. However, it is imperative that the SCR and all other exposed wiring leads be covered in some manner to prevent possible shock hazards.

Figure 4-52 shows how the SCR might be connected to three 115-volt receptacles to control three separate incandescent lamps or lamp strings. It must be remembered that this circuit, while operating from the AC line, has an output of direct current. Devices which are designed to operate only from alternating current will be damaged if connected to the DC output of this SCR control circuit.

Check-Out Procedure. Wire the circuit as previously indicated. Connect a DC load to the output of the SCR. This load should not exceed a 1000-watt rating, which will represent a current drain of a little less than 9 amperes. Turn R1 to the full counterclockwise position and insert the

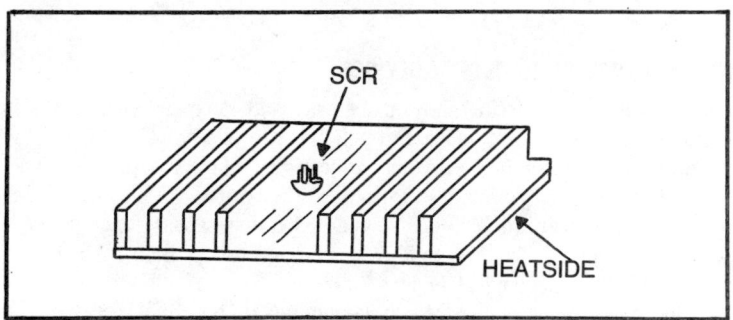

Fig. 4-51. Attachment of SCR to finned heat sink.

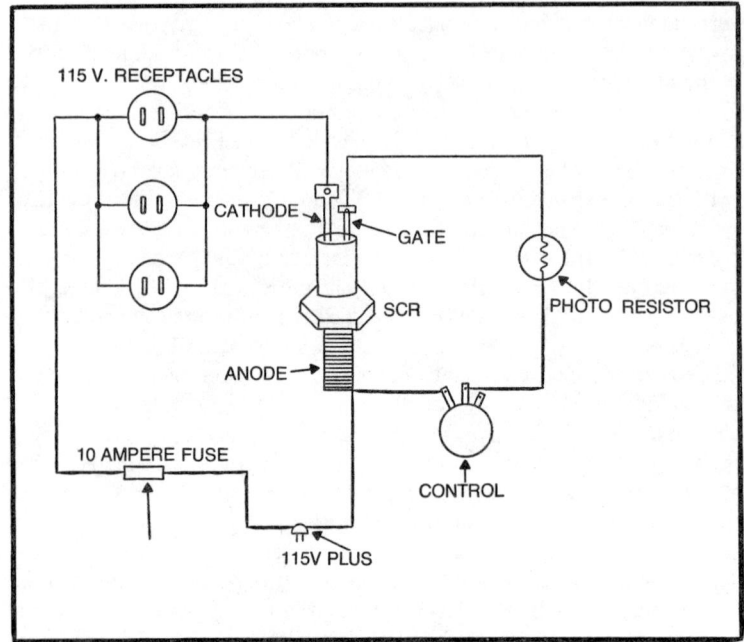

Fig. 4-52. Wiring of SCR circuit.

115-volt line plug into a wall outlet. Now, focus a light beam of the intensity which will be used for control purposes on the photoresistor. Adjust R1 in a clockwise direction until the load is fully operational. If an incandescent light serves as the load, the control is set when this lamp is at full brightness. Remove the light source from the photoresistor again will reactivate the load if the circuit is functioning properly.

If the circuit refuses to function, you may have a break in the wiring or the 10-ampere fuse may have opened. This is such a simple circuit that any problems should be easily traced; but if the circuit simply refuses to work and no obvious problem can be found, suspect a defective SCR and replace it.

HIGH-CURRENT CONTROL OF AC DEVICES

The previous circuit showed an easy and inexpensive circuit for controlling high-current loads through light intensity. The SCR is an excellent device for high-current applications, but its output is direct current which is not compatible with the majority of household appliances used today. In many applications, it will be found necessary to control AC devices using the direct method of conduction rather than resorting to relays and other mechanical devices.

Fortunately, a device which is similar to the SCR (electrically equivalent to two SCRs in reverse parallel) is the TRIAC. The TRIAC is available with high current ratings and may also be used to control a 1000-watt load. Its output is the same as its input, or alternating current.

Figure 4-53 shows a circuit which is identical to the previous SCR circuit except a TRIAC has been inserted in place of the SCR. The circuit is built in the same manner as before, but now transformers, motors, appliances, and other AC devices can be powered directly from the output of the TRIAC. R1 must be adjusted for full output voltage when PC1 is exposed to the light source. Measure the output voltage with an AC voltmeter. The voltage output can be varied through R1 and PC1.

This circuit is almost identical to the previous one, so all of the previous information on building, check-out, and troubleshooting will apply equally here. A line switch may be inserted between the plug and the fuse to allow the circuit to be completely deactivated without removing the plug from the wall.

SOLAR-POWERED RF OSCILLATOR

For those avid solar experimenters who would like to design a complex solar powered transmitter, electronic clock, or even a computer,

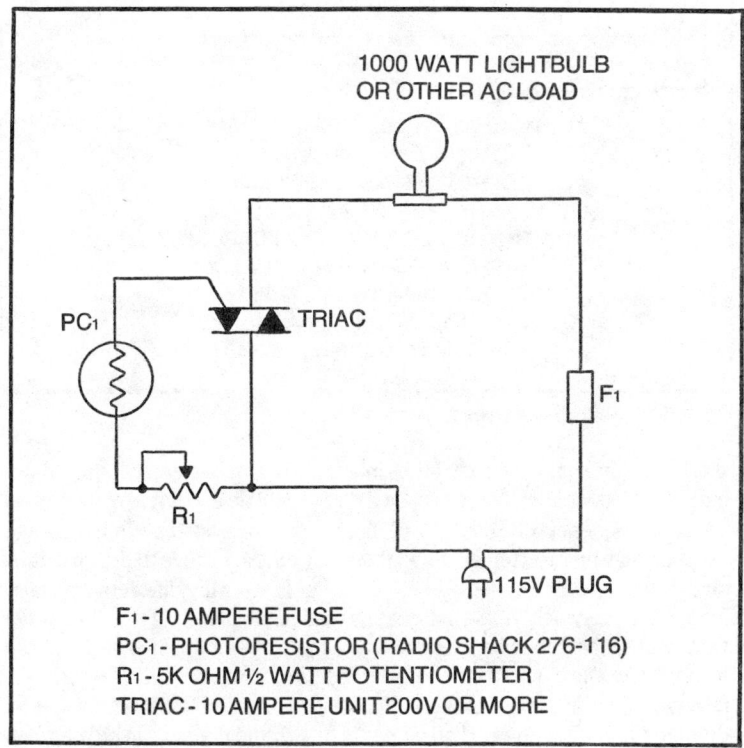

Fig. 4-53. TRIAC control circuit.

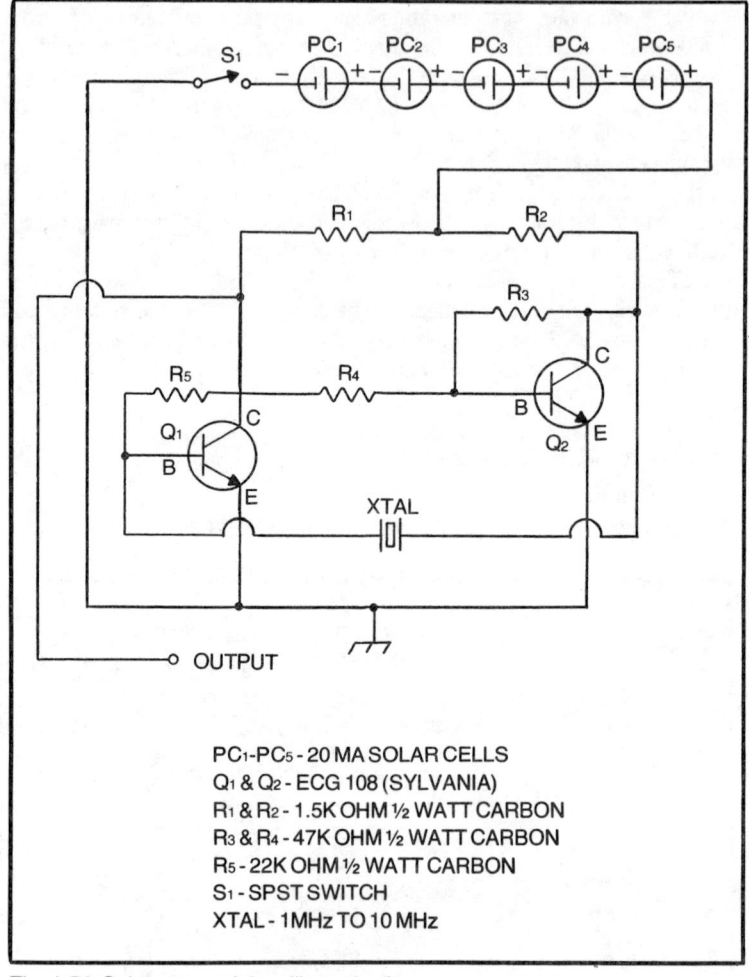

Fig. 4-54. Solar powered rf oscillator circuit.

the circuit in Fig. 4-54 may be of interest. It is a basic oscillator whose output is determined by the crystal frequency. The oscillator will function up to a frequency of about 10 MHz and contains no tuned circuits, so crystals may be inserted at will without any other changes to the circuit. A switch may even be added which can select from many different crystals. The output may be fed to a buffer stage and mixing circuit for a shortwave transmitter, or it may be connected to a logic circuit to act as a time base for an electronic clock. It may even be used in conjunction with a shortwave receiver as a crystal checker. The crystal is inserted and its output detected in the receiver. If no output is heard, the crystal can be assumed to be defective.

This oscillator is powered by five miniature solar cells which provide an output of about 2.25 volts DC. The circuit may be adequately driven by a bright room light and need not be in the bright sun. This is a miniature circuit and is assembled on a 2-inch square piece of perforated circuit board. Figure 4-55 shows the circuit board component layout, which has a balanced appearance. Two leads are brought off the circuit board to be connected to S1 and the solar cells. Another lead forms the output connection. The metal box in which this circuit is mounted serves as ground. The output lead should be shielded if it is not terminated in a shielded chassis connection.

Install the components as shown in Fig. 4-55. Cut all wires and component leads to minimum lengths. High-frequency oscillator circuits are no place for unduly long conductors which can cause feedback problems. This is the reason for the tiny circuit board section. All components may be placed in close proximity and minimum interconnection lengths are possible.

Figure 4-56 shows the final mounting configuration. An aluminum mini-box makes an ideal enclosure for this circuit board. It will be necessary to epoxy a piece of circuit board material to the top of the aluminum cover for mounting of the solar cells. The wired circuit board is mounted inside the box on standoff bolts. The solar cells are bonded to the perf board after wiring them in series. Two holes are drilled through the board and the aluminum case to which it is mounted to allow the positive and negative leads from the supply to pass to the switch and circuit board. The switch is mounted through the front panel.

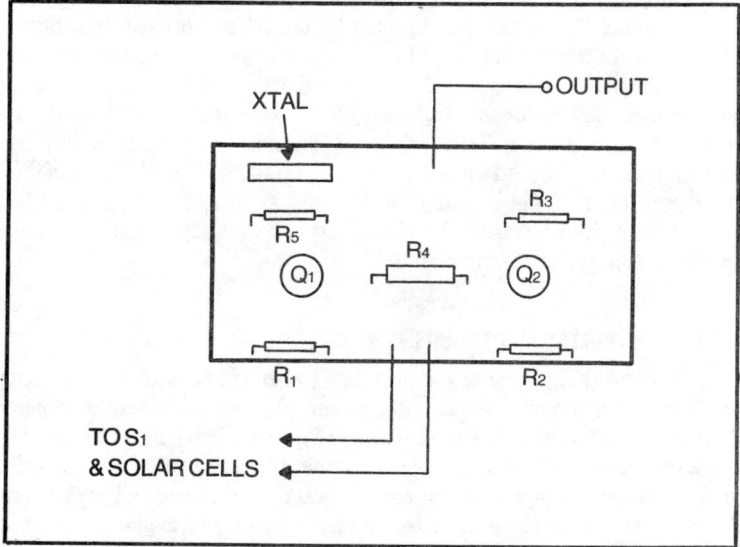

Fig. 4-55. Installation of components on circuit board.

Check-Out Procedure. After your circuit has been checked to make certain all wiring is proper, temporarily install it in the aluminum box. Do not tighten the mounting bolt nuts to the circuit board, as it may have to be removed should a problem develop. Solder the wire leads to the switch, solar cells, and chassis connector. This connector can be of any type which will mate the output of the oscillator to subsequent circuits with which it may be used. With the switch in the off position, shine a bright room light on the surface of the solar cells. Tune a shortwave receiver to the frequency of the crystal which has been inserted into its socket on the circuit board. Turn the switch to the on position and tune the receiver until the signal is heard. Make certain the receiver bfo is switched on so the continuous wave of the oscillator will be detected. A short length of hookup wire should be inserted in the output conector of the minibox and brought close to the receiver antenna terminals.

If output is not detected, try another crystal, one you know to be in working order. If a signal still cannot be detected at the receiver, measure the output voltage from the solar cells with a DC voltmeter. You should get a reading of just over 2 volts. An improperly connected solar cell could be the problem.

If the circuit still won't operate, remove the circuit board. There must be a wiring error or a defective component somewhere. Check each component individually and replace any which are questionable. Reconnect the board and try again. Since the output of this oscillator is very low, it may be necessary to place the output lead even closer to the receiver antenna terminals, especially if the receiver sensitivity is not very good.

Once the circuit has been checked out, tighten the circuit board to the mounting bolts and enclose the entire unit. This project has many applications, both as a building block for more elaborate solar circuits and for instantaneous checking of crystal activity. If used for the latter purpose, it may be desirable to install a crystal socket on top of the aluminum case and run wires to the circuit board. Then, the crystal to be checked may be inserted without having to gain access to the interior of the aluminum case. If the circuit is to be used as a time base, it may be desirable to do away with the crystal socket completely and wire the crystal directly to the electronic circuit.

LIGHT-CONTROLLED AUTOMOBILE FINDER

How many times have you been to a sports arena, outdoor event, or even a large shopping center and had difficulty finding your automobile because you couldn't remember exactly where you had parked it? This is not a tremendously serious problem, but the circuit shown in Fig. 4-57 will make it easier for you to locate your car after the sun goes down. All you have to do when emerging from that stadium or shopping mall is look in the general direction where you parked your car and locate the automobile with the flashing light sitting on the rear deck.

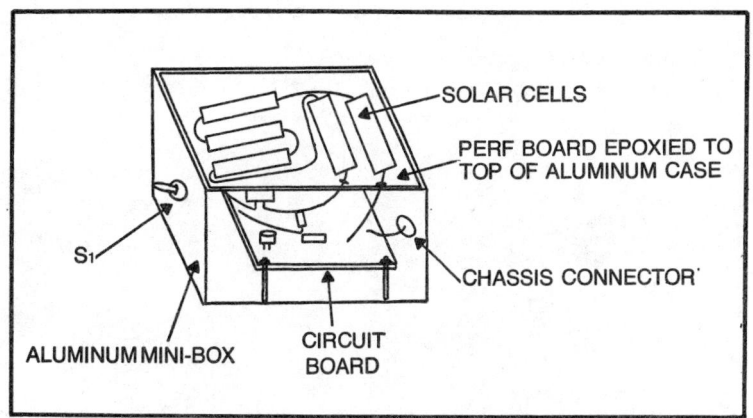

Fig. 4-56. Final mounting configuration in aluminum mini-box.

This circuit is a light flasher which is controlled by the amount of light present. The circuit is not activated until the sun goes down; then PC1 stops sending current through the base of Q1 which reverts to its non-conducting state. This causes K1, which has been held closed, to become de-energized, activating the circuit powered by B1 through its normally closed contacts. This causes the two 6-volt lamps to glow. One lamp is located on the outside of the plastic enclosure for visibility, but the other is located in close proximity to PC2, a photoresistor. When this lamp reaches full intensity in a fraction of a second, it causes Q2 to conduct, activating K2. When this second relay is energized, it breaks the circuit to the lamps. Both lamps go out, but as soon as they do, PC2 returns to a high resistance state and Q2 goes non-conductive. This de-energizes K2 and the lights come on again. The whole process is repeated about every second. Whenever Q2 turns on, it also turns on K1, which turns off the lamps. This action in turn causes Q2 to turn off and the lamps come on again. This is the complete flashing cycle. All PC1 and Q1 do is activate the flashing circuit when the sun goes down. The entire circuit, with the exception of the 6-volt lamp, is operated from your automobile battery. It draws nominal current, so battery discharge is no real concern. If you prefer to use 12-volt lamps, you may even do away with B1 and replace it with connection to the positive and negative terminals of the automobile battery.

This circuit has many other uses around the home and workbench. It can be used to turn other circuits on and off or even as a burglar alarm which is visible rather than audible. It would be possible to build a flashing circuit using solid-state components alone, but the cycling photocell version is more pertinent to the discussion in this book.

Figure 4-58 shows the parts layout on a small section of perforated circuit board. The two relays are the largest components which are mounted here. R3 is mounted through the board and adjusted by a

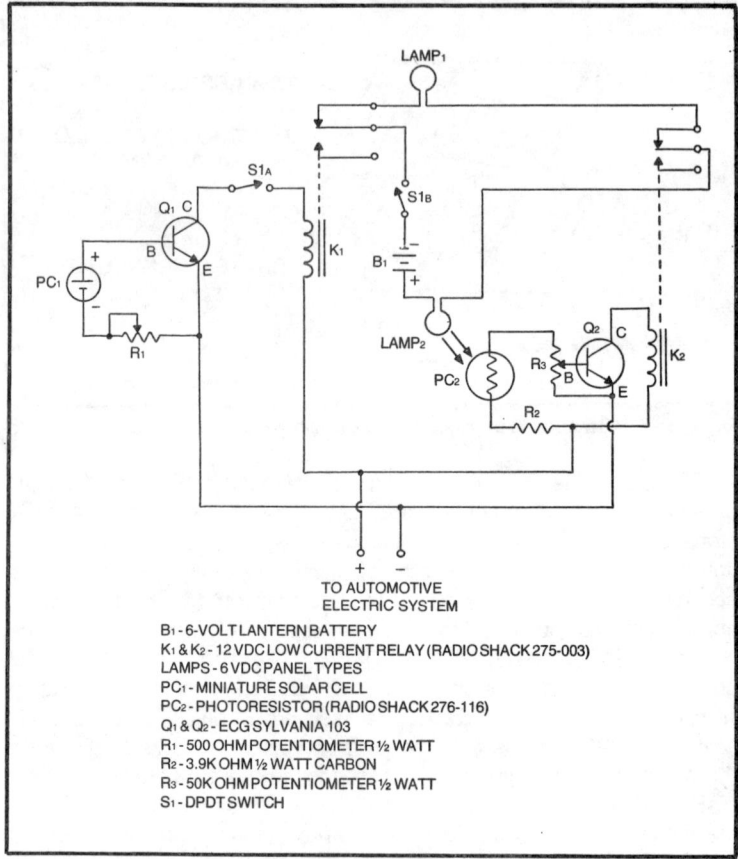

Fig. 4-57. Light-controlled flasher circuit.

screwdriver during the set-up procedure. It is then left in place and need not be adjusted again. The other variable control, R1, is brought out to the enclosure for external adjustment.

Notice that one of the 6-volt lamps is mounted directly to the circuit board near PC2. It will take some adjusting of the distance and angles between these two components in order for the flashing circuit to work as desired. By increasing the distance, the external lamp will flash more slowly. Decreasing the distance will bring about an opposite effect.

Figure 4-59 shows a suggested mounting configuration in a plastic enclosure. If this case is large enough, you may even be able to install B1 within its interior. B1 is a 6-volt lantern battery which should last for many hours of operation if small lamps are chosen. The external lamp may be a larger type than the one which is mounted to the circuit board. Only a small amount of light is needed to trigger PC2. A brighter light might be desirable for the case-mounted position to increase visibility. The easiest

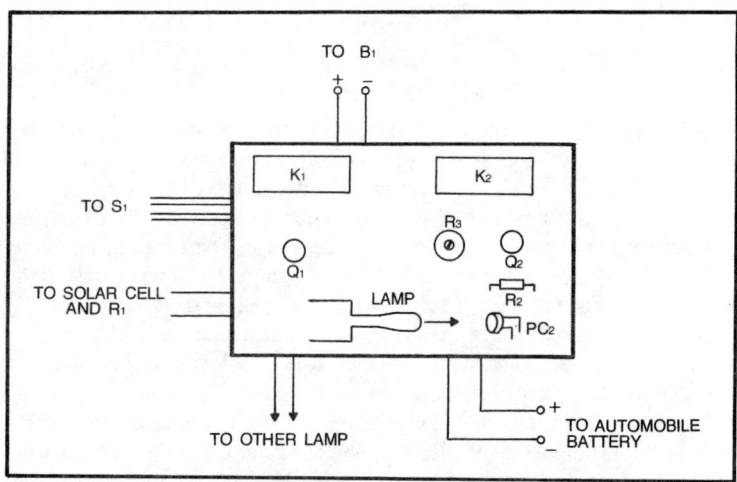

Fig. 4-58. Parts layout for flasher on circuit board.

way to use this circuit is to provide a long power cord fitted with a plug which will insert in the cigarette lighter well of your automobile. The circuit may then be placed on the rear deck inside the back window, the plug inserted, and S1 thrown to the on position. During the daylight hours, PC1 will be providing current to Q1 and the external lamp will not flash; but when the sun begins to go down, at a point determined by the setting of

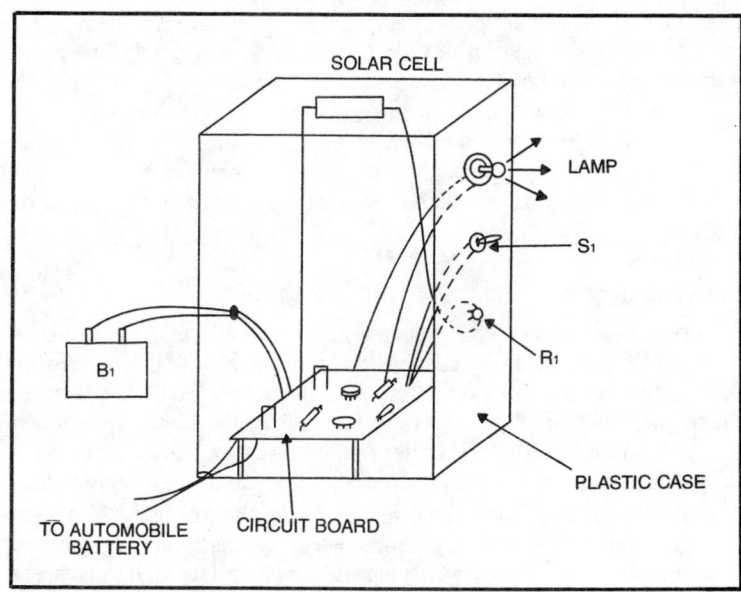

Fig. 4-59. Mounting in plastic enclosure.

145

R1, the flashing circuit will be activated. The solar cell is mounted atop the plastic case and samples the ambient light.

Check-Out Procedure. After the circuit board has been wired, install it in a plastic case and make all connections to the chassis-mounted components, B1, and to the automobile battery. Make certain S1 is in the off position. (During the test, any 12-volt power supply will suffice.) After you are certain that all connections have been properly made, turn R1 fully counterclockwise. Shine a bright source of light on PC1 and turn S1 to the on position. You may hear K1 click as it is energized and there may even be a flash from the external lamp, but this is all that should occur. If the lamps glow or even flash, readjust R1 until they stop. If R1 was originally set in a full counterclockwise position, this latter condition should not occur. Now, direct the circuit board lamp onto PC2. A piece of tape may be used to hold both components in place. Cover the solar cell with your hand and adjust R1 until the lamp glows. You will also hear K1 click again as it becomes de-energized. As soon as the lamps come on, they should begin to flash. If they remain on, adjust R3 until the flashing action begins. If the circuit still won't flash, couple the lamp and PC2 more closely. If they flash too rapidly, separate these two components by a greater distance.

This is a fairly complex circuit, so make certain all your wiring connections are correct the first time. Make certain B1 is fully charged and will operate the circuit.

Once the correct setting for the lamp and PC2 are found, these components may be permanently attached to the circuit board by bonding them with epoxy cement. Alternately, you may wish to simply tape these items into place to allow for future flash-rate changes.

Once it has been established that the circuit is operational, you can adjust the sensitivity to ambient light through R1. This circuit can be set so that it begins flashing in dim light or in darkness, depending on where R1 is set. R3 can be used to adjust the sensitivity of PC2 to the circuit board lamp.

LIGHT MODULATOR

This next project does not use any photocells or light-sensitive devices. Rather, it is designed to provide a source of power or control for some of the devices already discussed in this chapter. When purchased on the retail market, similar circuits are often called light organs. They attach to your stereo and drive a multitude of colored lights which are in sync with the music or other audio information produced by the stereo system.

The circuit in Fig. 4-60 is a light modulator. Audio from one system is transposed into light variation in another. The controlling device is a 1-ampere SCR which will provide modulation control for an incandescent lamp or any other DC load up to 100 watts.

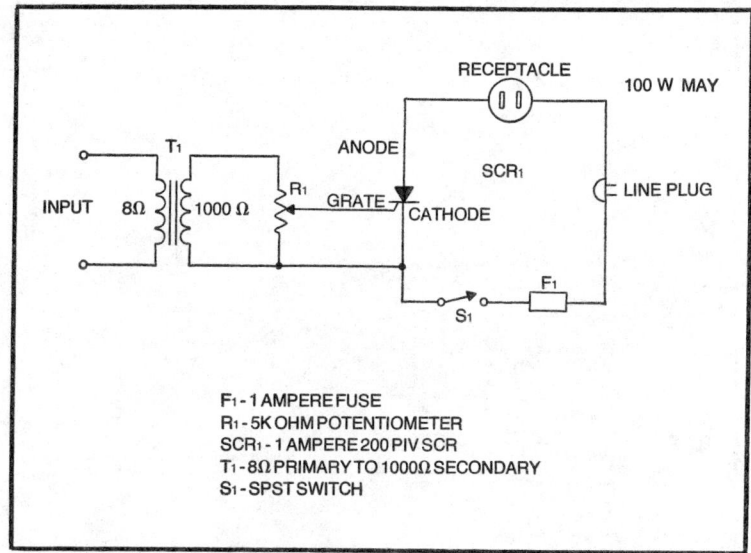

Fig. 4-60. Light modulator circuit.

The circuit is quite simple and this low-powered SCR will not require an external heat sink. It may be mounted directly to the perforated circuit board along with the audio transformer and fuse. The audio output from a speaker is connected to this circuit through T1. The input leads to this transformer are connected across the speaker leads. The secondary feeds these pulses between the gate and cathode of the SCR. R1 controls the amount of drive these two SCR elements obtain. This drive controls the conduction of the SCR. Loud audio signals will cause it to conduct fully, while softer signals will bring about only partial conduction. All of these conduction angles have a direct effect on the intensity of the incandescent lamp attached to the circuit through the receptacle.

Figure 4-61 shows the component placement on the circuit board. The inset shows how R1 is to be wired. Three leads from the transformer and SCR connect to R1, which is chassis-mounted. Other leads from this board connect to the power plug and receptacle and to the input terminals where the speaker leads are attached. It should be possible to mount all of the components on a 2-inch square piece of perf board, providing that a miniature audio transformer can be obtained.

Check-Out Procedure. Referring to Fig. 4-62, the circuit board is mounted inside an aluminum mini-box. Make all connections to the circuit board, receptacle, input terminal, R1, and S1. Make certain S1 is in the off position before inserting the line plug into the wall. Once this is done, rotate R1 until the lamp glows. If the lamp lights, this means the circuit is operational. If not, F1 is blown, S1 is not closed, or the SCR is defective. Now, back off on R1 until the lamp dims and then goes completely out.

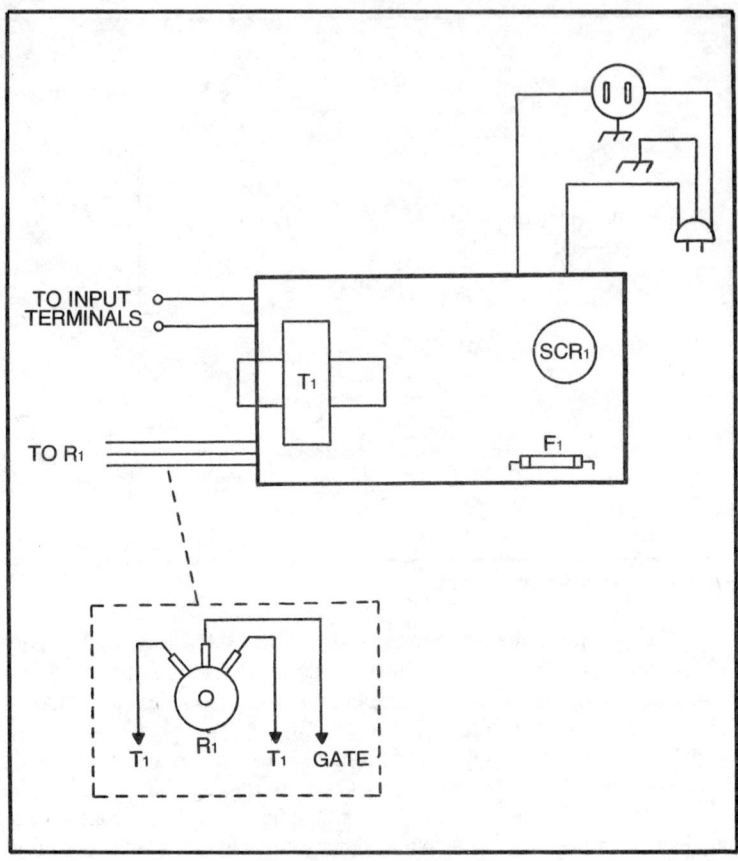

Fig. 4-61. Light modulator components on perforated circuit board.

Connect the speaker leads to the input terminal and tune in a radio station on your stereo receiver, or play a tape or record. The light should come on again. It may come on and remain at one brightness level regardless of the loudness of the music. Back off on the control even further until the light intensity begins to follow the rhythm and volume variations of the music. Again, R1 is your sensitivity control. It will allow you to adjust the light modulator's reaction to different volume levels.

This circuit can be used with signal generators, audio amplifiers, and other sources of audio output to provide varying light intensity to light sensitive devices. With this modulator, it is possible to pulse light sensitive circuits by pulsing the audio information which is fed to the modulator input.

This circuit converts audio information into video information. Some of the electronic circuits already discussed can convert video information back into sound, into voltage pulses, and even into metering pulses.

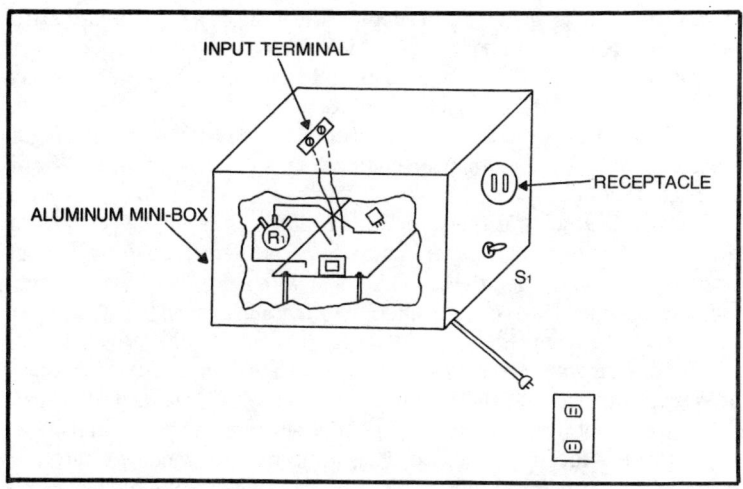

Fig. 4-62. Mounting of circuit board and other components in aluminum mini-box.

MOTORIZED PINWHEEL

This project is one of the few which has no known uses. It is offered purely for the pleasure of the builder, as it is one of the simplest circuits there is to put together. It's guaranteed to work every time and it takes less than fifteen minutes to complete. It's the solar-powered pinwheel shown in Fig. 4-63.

All it does is spin whenever sun strikes the surface of a single solar cell, which is its only power source. Referring to the pictorial drawing, a

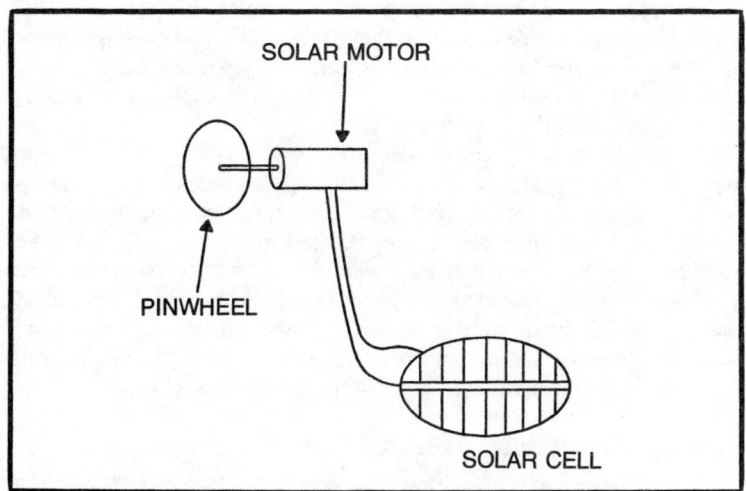

Fig. 4-63. Motorized pinwheel circuit.

149

miniature motor is mounted to a support which can be a ruler, a cardboard box, or just about anything through which a hole can be punched to insert the motor frame or which can be bonded to the motor frame with epoxy cement. The motor is a specialized component but is available from Edmund Scientific Company for less than $5. It will operate from a single solar cell with a 125-milliampere output. This cell is also offered from Edmund Scientific for less than $5.

Mount the motor any way you want to and cut a circular wheel from a piece of lightweight, stiff paper. This wheel can be painted in any design you wish and is attached to the motor shaft with a drop of epoxy cement. Once the cement has set, the wheel will not slip around the shaft as it turns. It will turn with the shaft. All it takes to complete this project is two conductors carrying power from the solar cell. When these connections are made, a bright light played upon the surface of the solar cell will cause the motor to start turning. When the light disappears, the motor stops.

The ideal location for this useless project is in a window where the solar cell may catch the rays of the sun. The motor will slow down and speed up with the intensity of the available light.

Perhaps we are being a little bit too hard on this project, as it will entertain adults and children for long periods of time and is guaranteed to be a conversation item when visitors to your home witness its operation. Bearing this in mind, it may not be fair to say that this is a completely useless project—but almost.

Going a step further, the avid experimenter might wish to install this motor in a tiny slot racer or other type of toy vehicle. With the solar cell mounted on top, an interesting solar toy might be developed. If three solar cells are used, only 30 milliamperes of current are needed to turn the motor. The current increases as the voltage drops. Edmund Scientific Company states that the motor can be operated from as little as 0.35 VDC, which is a tenth of a volt less than most solar cells produce. Any device which presently uses a small DC motor might be converted to solar power by using the motor specified for this project and the number of solar cells required to obtain the proper drive.

If you think hard enough, you may come up with many different applications for a solar powered motor. It might be necessary to decrease the speed through tiny reduction gears, but this should be of no great concern to the budding electro-mechanical experimenter. Certain cam arrangements could be worked out whereby all kinds of miniature work functions could be performed. But back to the project at hand. If you don't have a talent for electro-mechanics, you can always sit by your window and watch the pinwheel turn around and around and around. At least you can say you are continuing to experiment with solar-powered circuits.

LIGHT-REGULATED POWER SUPPLY

The project in Fig. 4-64 provides an exercise in varying the output voltage of a power supply which is powered from the 115-volt house

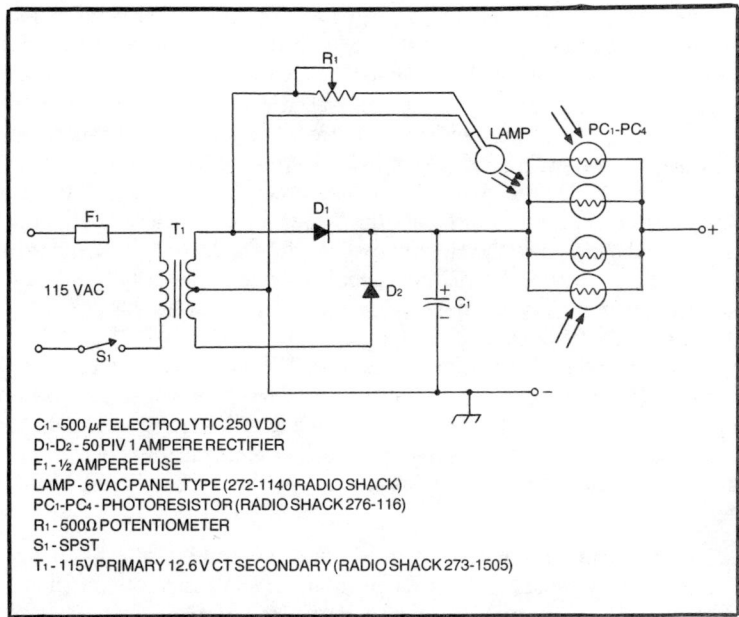

Fig. 4-64. Light-regulated power supply circuit.

current line through light intensity. This project provides a stable DC output for powering transistorized circuits and at the same time controls voltage output by the intensity of the light which is mounted to the power supply circuit board.

Be referring to the circuit diagram, it can be seen that the secret of this light-controlled circuit is four photoresistors, which are placed in parallel at the output of the supply and in series with the light bulb and its power source. When the light is increased in intensity, the resistance of the photoresistors decreases and less voltage is dropped across these components. When the light is dimmed, the resistance increases and the voltage is lowered or cut off completely.

The panel lamp is a 6-volt type and is powered from the secondary of the transformer before it is rectified. This circuit only drops voltage when the supply is furnishing power to an electronic load. It is the current through the photoresistors which causes the voltage to be dropped.

This circuit uses a very common transformer which has a 115-volt primary and a 12.6-volt secondary. The secondary also has a center-tapped connection. A full-wave center-tapped rectifier circuit is used, along with a 500-microfarad capacitor. Voltage measured at the capacitor will be around 11 volts when the power supply is under no load, but by controlling the amoung of light which is directed to the treated surface of the photoresistors, the output voltage to the device under power will be varied as desired. When the light is turned to a very dim or off state, no current will flow to the load and voltage will be zero. As the brightness is increased, the voltage level will begin to rise and can be adjusted to a value of nearly 9 volts under conditions of extremely low current drain. This is an unusual way of controlling voltage output and is not all that practical, but it does demonstrate the effect of resistance, photoelectric resistance, in this case, on current flow and voltage in an electronic circuit.

The 6-volt lamp is installed on the circuit board near the three photoresistors in a similar fashion to the earlier automobile locator circuit. Figure 4-65 shows how the circuit board is wired with the two rectifiers, the capacitor, photoresistors, and lamp. It may be necessary to cant the photoresistors slightly in order to allow them all to receive equal amounts of light from the 6-volt lamp.

Figure 4-66 shows how the circuit board is installed in an aluminum mini-box. T1 is mounted at the bottom of the box with its connections from the line cord and fuse to the primary winding and from the secondary winding to the circuit board. The center-tapped lead from the transformer is connected to the aluminum chassis, as well as to the circuit board contact.

Check-Out Procedure. With S1 in the off position, insert the line plug into the receptacle. Turn R1 to its maximum resistance position, which should be fully counterclockwise. Connect the probes of a voltmeter across the positive and negative output terminals. Turn S1 to the on position. The voltmeter should indicate about 10 volts DC. Now, connect an electronic load to the circuit. A 6-volt panel light which draws about 100 milliamperes will be ideal. Once a load has been established, switch on the supply again. If a light bulb is used, it should glow brightly. Turn R1 slowly clockwise. The light bulb load at the output of the power supply should begin to dim. If another type of load is used, a voltmeter still placed across the output should begin to indicate a drop in voltage. If no drop is indicated, couple the internal lamp more closely to the photoresistors.

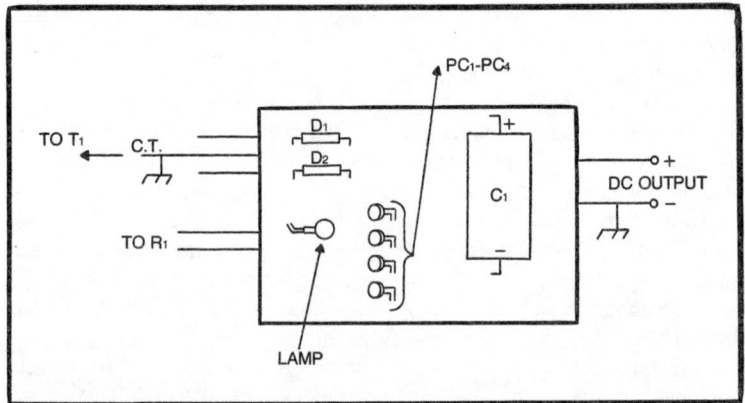

Fig. 4-65. Component placement for power supply.

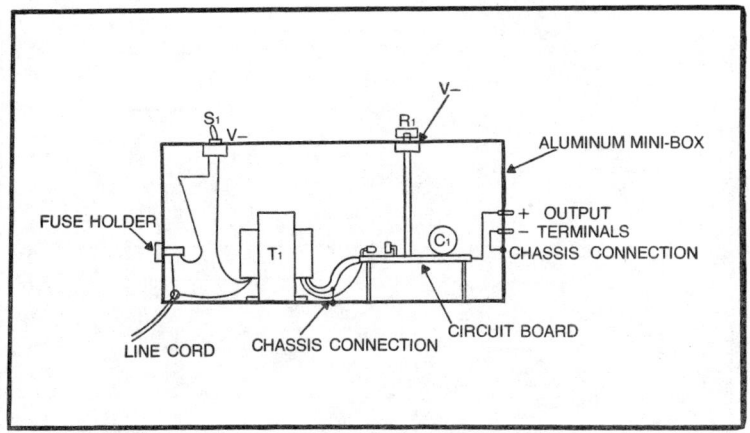

Fig. 4-66. Mounting of power supply circuit in enclosure.

If the circuit should refuse to work and no output voltage is indicated during the initial testing stages, examine the circuit board-mounted lamp. With R1 in the minimum position, it should glow. If it does, the problem will be in the rectifiers, capacitor, or photoresistors. The first two items are polarized devices and must be connected to the proper circuit contact. A reversal here could cause the supply not to operate at all.

On the other hand, if the lamp refuses to glow, your transformer is not working. This may be due to a blown line fuse (F1) or a break between the line cord and receptacle. T1 could also be defective. Make certain that the power receptacle is supplying operating current by plugging in a lamp or other device which is known to be working. Also, look for broken wires around the switch and fuse holder.

Once your circuit is made operational, it may be used to power simple electronic circuits which do not require a high degree of regulation. Some solid-state projects using integrated circuits or transistors will not operate properly due to the fluctuation of output voltage with varying amounts of current demanded by the load.

REMOTE LIGHT-CONTROLLED POTENTIOMETER

An earlier project described a way to interrupt the sound of your television receiver during a commercial. This project will allow you to be able to control the volume setting on a nearby television, stereo, or radio receiver without ever leaving your chair. This is not a complex circuit electronically, but some fairly intricate relay wiring is required to be able to control the potentiometer in both directions for raising and lowering volume.

The circuit is shown in Fig. 4-67. In many ways, it is a carbon copy of an earlier project, the solar cell switch. As a matter of fact, it is two carbon copies, as duplicate circuits are used to control two relays. The relays are

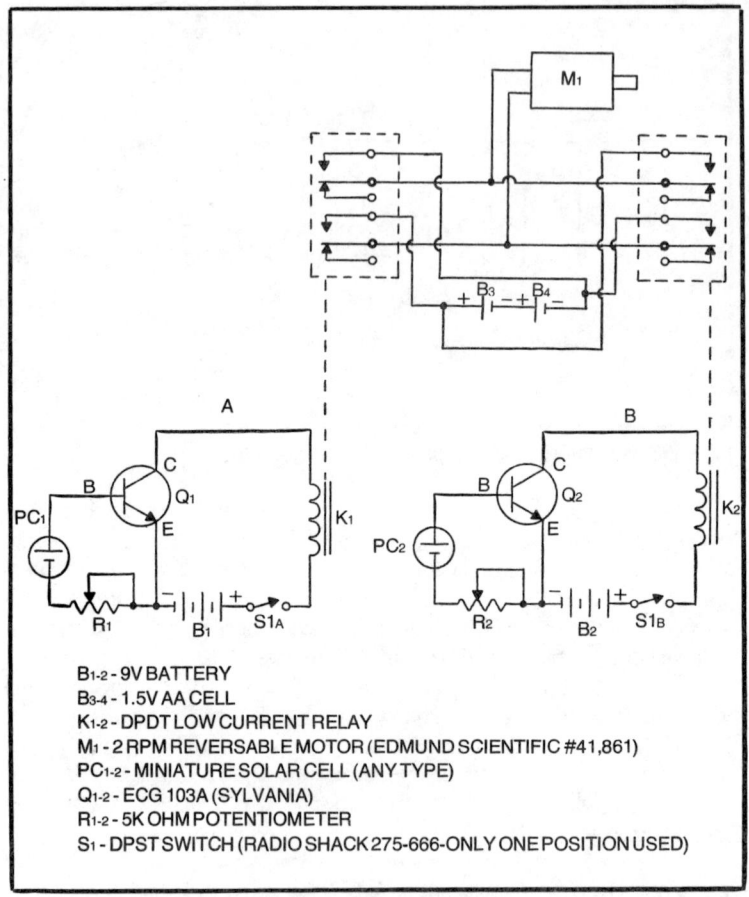

Fig. 4-67. Remote light-controlled potentiometer circuit.

a bit different. They are DPDT types although a DPST type would be preferred. This latter type is more difficult to find, so the common DPDT version is used in a DPST configuration. Half the contacts of each relay are not used.

The heart of the circuit is M1, a two-rpm miniature motor, which is operated from a 3 VDC power source provided by two AA cells in series. The motor draws less than 15 milliamperes, so the batteries should last for quite some time. The electronic portion of the circuit is powered by two 9-volt transistor radio batteries, although one may be used by paralleling the power connections. Using two batteries gives the circuit a longer operational life.

The miniature motor is reversible. Circuit A causes the shaft to turn in one direction, while Circuit B will cause it to reverse itself. When a light beam is played upon the surface of PC1 in Circuit A, K1 closes its contacts and power is supplied to the motor. The shaft begins to turn and continues turning as long as the light beam stays on PC1. To control the volume in the other direction, the light beam is removed from PC1 and directed onto the surface of PC2. Initially, the motor stops because K1 opened its contacts when the light beam was removed from PC1. When it strikes PC2, K2 closes its contacts. This supplies power to M1 again, but this time the polarity is reversed. The motor shaft turns in the opposite direction. It is important never to direct a light beam on K1 and K2 simultaneously, as the battery terminals will be shorted. R1 and R2 control the sensitivity of each circuit to light. These should be adjusted so that only a focused beam of light will trigger the relays. When sensitivitiy is improperly set, normal ambient lighting conditions could cause both circuits to trigger simultaneously.

Figure 4-68 shows the component placement on the circuit board. Only the relays, transistors, and circuit batteries are located here, so the board need not be very big. The real complexity develops when the circuit board is placed in a plastic or aluminum box and must be wired up to the chassis-mounted components. Figure 4-69A shows the front panel of this box. PC1 is located at one end, while PC2 is at the other. This prevents both circuits from being triggered simultaneously by a single light source. The solar cell variable controls are located below and above each unit, respectively. S1, located in the center of the front face, is a DPST switch which controls both electronic circuits.

Figure 4-69B shows a side view of the same enclosure. The circuit board is mounted on stand-off bolts near the back wall. The "AA" cells which power the motor are mounted in a Radio Shack holder and secured beneath the circuit board. The relay connections are made between the

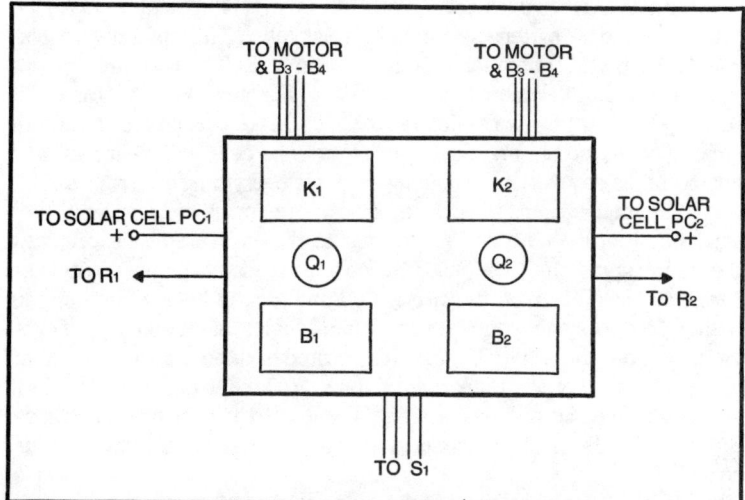

Fig. 4-68. Component placement on circuit board for potentiometer control.

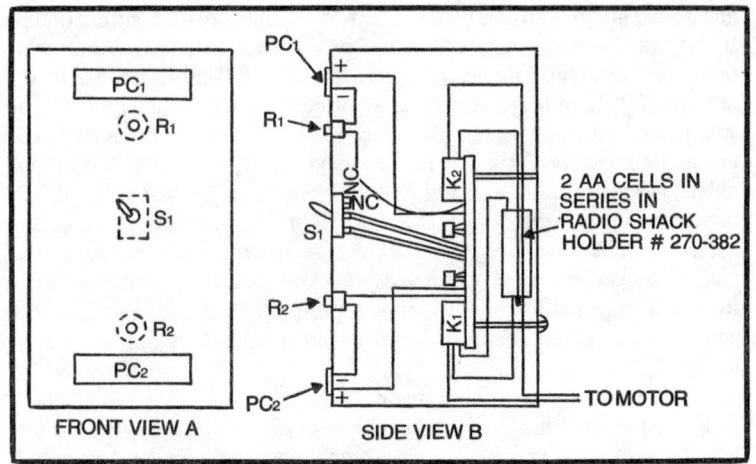

Fig. 4-69. Mounting of components and circuit board in plastic enclosure.

motor and battery. Motor wiring is brought out of the case in a two-conductor cable. Make certain you wire the relay connections to the battery, as shown in Fig. 4-67. Make the other connections as indicated.

Figure 4-70 shows the mounting of M1 to the front panel of the device it is to control. The motor specified has mounting brackets which will allow it to be secured as shown in the drawing. A coupling between the motor shaft and potentiometer shaft can be fabricated from an aluminum crimp connector or even from cellophane tape in a pinch.

Check-Out Procecure. With S1 in the off position, make all connections to the batteries, motor, and solar cells. The motor need not be connected to the potentiometer at this time. Set the circuit and its enclosure at the site where it is to be more or less permanently located and adjust R1 and R2 for maximum resistance. Make certain that the room lighting is not too bright. Cover one of the solar cells with your hand and turn S1 to the on position. If nothing happens, everything is all right so far. But if the motor begins to operate, there is too much ambient light. Try adjusting the variable control connected to the uncovered solar cell until the motor stops. If it continues to run regardless of the position of the variable control, turn off the circuit. R1 and R2 will have to be replaced with higher resistance control units. This assumes that the amount of light present in the area where the circuit is located is of normal intensity, that which is usually present most of the time. You might replace R1 and R2 with 25- or even 50-Kohm controls. If the sensitivity adjustment becomes too critical with these high values, a slightly lower resistance might be considered.

During the initial testing, if the motor does not activate, the values of R1 and R2 are alright. You may now uncover the other solar cell. The motor should still remain in an off state. Now, adjust one of the controls until the motor begins to turn. Back off on the control considerably until the motor stops again. Adjust the other control in the same manner.

Using a narrow-beam flashlight, direct the light ray on one of the solar cells. The motor should begin to operate. If it does not, back off on both controls a bit more, as both relays may be triggering simultaneously. Direct the ray onto one solar cell again. When the motor begins to turn, turn off the flashlight. The motor should stop. Direct the flashlight to the other solar cell and repeat this testing process.

If the motor turns in the same direction at all times, regardless of which cell is activated, you most likely have a wiring error between the relay contacts, B3 and B4, and the motor leads. If the motor refuses to operate at all, check to see if K1 and K2 are triggering. A click will be heard when they activate. If nothing is heard, B1 or B2 may be dead or a wiring error has occurred. Also, check B3 and B4 often to make certain they are fresh.

Once the circuit has been made operational, the motor may be connected to almost any type of volume control for remote activation. It is interesting to note that the reversable motor may also be used with this circuit to tune a radio by connecting it to the tuning shaft. Several of these motors could be used to remotely control the volume, frequency response, and tuning of a stereo receiver.

LIGHT-CONTROLLED HOLDING SWITCH

All of the previous light-controlled switching circuits have been of the momentary type; that is, to accomplish a switching function, a light source always had to be present. When the light source was removed, the switch returned to its original position. A device could be turned on only so long as there was a controlling light source present. If we wanted to use one of

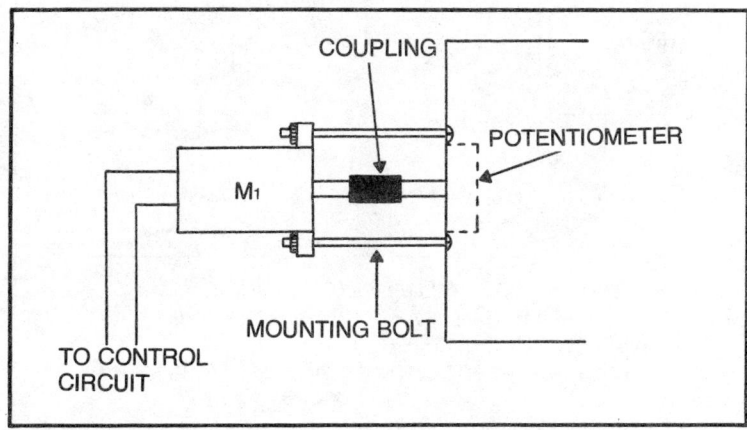

Fig. 4-70. Attachment of miniature motor to potentiometer shaft.

these switches to turn on a television receiver, we would have to keep the light beam in contact with the solar cell or photoresistor for as long as we wanted the set to be operational. This is impractical and inconvenient. Figure 4-71 shows a relay control circuit which may be connected to any of these former switching projects and will do away with the inconvenience discussed.

Relay K1 is a 115-volt AC type whose operation is controlled by the relay in the light-activated switch. When this latter relay closes, power is fed to K1. K1 is not a common type of relay. It's known as a ratchet relay and requires pulses of current to be operated. When 115 VAC is applied to the coil, the relay contacts are switched to the on position. But when this power is removed, the contacts still remain closed. When 115VAC is applied again, the contacts open up and remain open until power is removed and then applied again. Pulses of current determine which state the relay contacts are in. It can be seen that it is not necessary to furnish a continuous supply of operating current. Using this system, a controlling light beam need only be applied to the photosensitive device for a moment to initiate triggering action. Using some of the earlier switching circuits described, a light beam is applied to a solar cell in the base leg of the transistor wiring. This causes its relay to close contacts. This action feeds

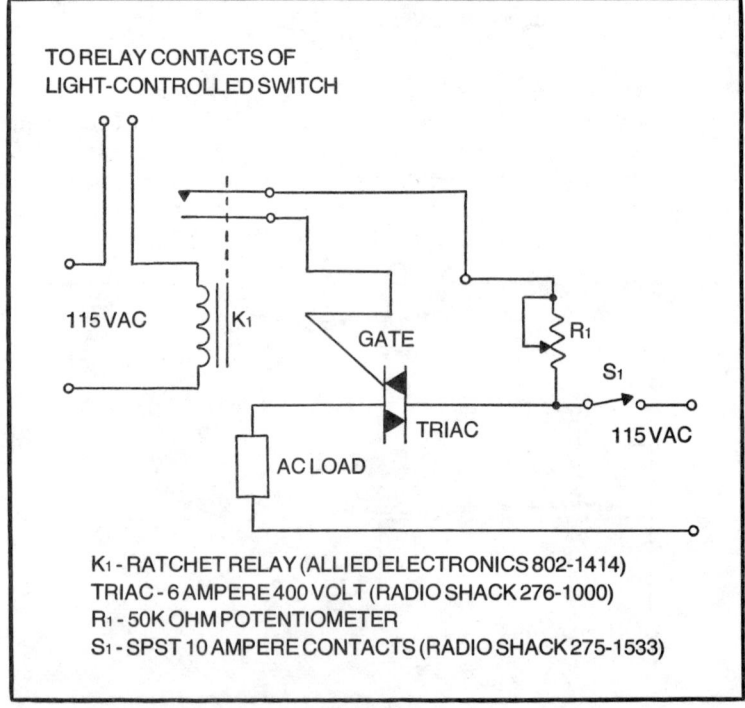

Fig. 4-71. Light-controlled holding switch circuit.

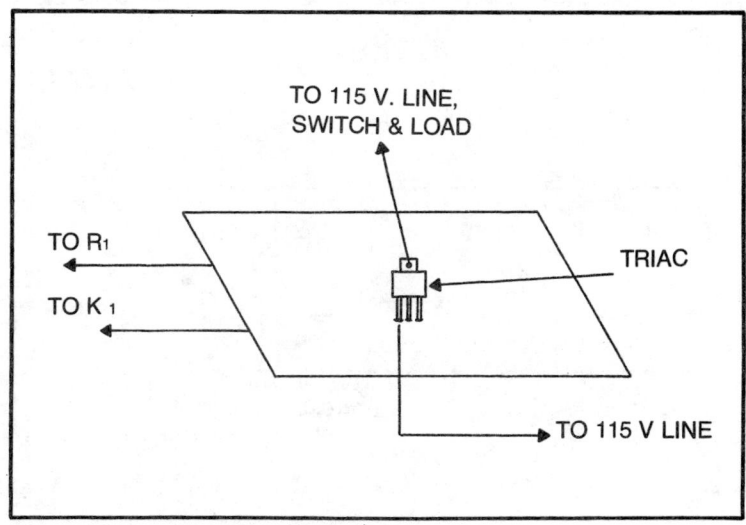

Fig. 4-72. Mounting of TRIAC on circuit board.

power to K1 in Fig. 4-71, causing its contacts to close. When the light source is removed from the solar cell, the first relay will open its contacts again, but K1's contacts remain closed. When the light source strikes the solar cell again, the first relay closes. This sends another pulse of current to K1, which now opens its contacts and keeps them in that state until the first relay opens and then closes again.

Using the circuit shown in Fig. 4-71 with any of the previous switching projects with normally opened contacts, on-off control of various electronic devices can be had with momentary pulses of light. Referring to this drawing, K1's contacts have been connected to a TRIAC control circuit. This avoids the direct switching of medium current AC lines, which can cause contact pitting and burning. R1 is adjusted so that the TRIAC will conduct alternating current when the relay contacts are closed. When they open, the TRIAC goes to a non-conductive state and current is removed from the AC load.

A 6-ampere 400-volt TRIAC is used in this circuit because it is a common unit easily obtained from local hobby stores. S1 is the on-off switch controlling the 115-volt AC supply. Normally, K1 will receive its operating current from this same source. With the TRIACs indicated, this circuit can handle up to a 650-watt AC load. The K1 contacts need only be rated at a half ampere or so because very little current flows between the gate and R1.

Figure 4-72 shows the TRIAC connections. This solid-state device is the only component on a very small piece of perforated circuit board. Figure 4-73 shows how it is mounted in a plastic or aluminum box along with K1, the light-activated switching circuit, and the other controlling components.

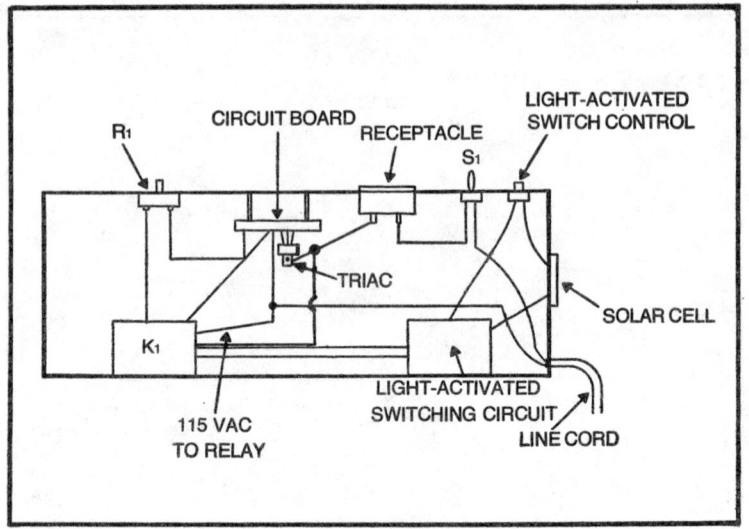

Fig. 4-73. Mounting of components in enclosure.

Check-Out Procedure. This check-out procedure assumes that the light-activated switching circuit has already been built and is fully operational. With S1 in the off position, adjust the light-activated circuit for proper triggering by listening to the relay click when a light source is directed upon the solar cell. Once this circuit is adjusted, turn S1 to the on position with a load connected to the receptacle. If this load is activated, back off on R1 until it ceases to operate. Chances are R1 will be left in a maximum-resistance position. Now, direct the light beam on the solar cell; the load should be activated again. If not, adjust R1 until power is delivered. Turn off the light beam. The load should remain activated. To remove power from the load, the solar cell is exposed to the beam once more. The load should cease to function.

This is a very simple circuit and should operate on the first try, providing that the light-activated switch is functioning properly. If the problem seems to be in this switch, refer to the check-out procedure provided with that project for troubleshooting information. The things which could cause the portion of the circuit shown in Fig. 4-71 to malfunction would be a defective relay or TRIAC, or a broken wiring connection.

LIGHT-POWERED FIELD-STRENGTH METER

It is very easy to obtain a good reading on rf-powered field-strength meters from transmitters with outputs of 500 or 1000 watts; but when it comes to CB radios, walkie-talkies, and low-powered oscillators, it's a different story completely. Some form of amplification is needed in order for these low-powered signals to provide adequate current to drive the meter. Since it is not convenient to amplify the power output of these small transmitting devices, an amplifier must be built into the field-strength meter circuit. The ideal thing about most field-strength meters which do not use this amplification is the fact that the power from the transmitter alone is all that's needed to operate the instruments. No batteries are needed.

The circuit shown in Fig. 4-74 has all of the advantages of an rf-powered field-strength meter, plus it has a built-in amplifier which will provide adequate meter readings from the outputs of even the smallest transmitter. It uses no batteries because it is powered entirely from the sun (or other source of light).

By referring to the schematic, you can see that only two solid-state devices are used, a germanium diode and a bipolar transistor. Many different types of transistors will work in this circuit; just make certain that if a substitute is chosen, it is of the n-p-n variety. P-n-p transistors may also be used, but it will be necessary to reverse the connections of the diode, meter, and solar cell string. The meter is an inexpensive 0/1 milliampere type, although other ranges will also work, with 10 milliamperes being the highest value desired. If smaller meter values are used, it may be necessary to increase the resistance of the variable control for smoother tuning. A small antenna serves to pick up the radio frequency transmission and can be almost any convenient length, from 10 inches upward.

Figure 4-75 shows the component installation on a small piece of perforated circuit board. Field-strength meters are traditionally small instruments, so the circuit board should be as tiny as possible. Wiring is

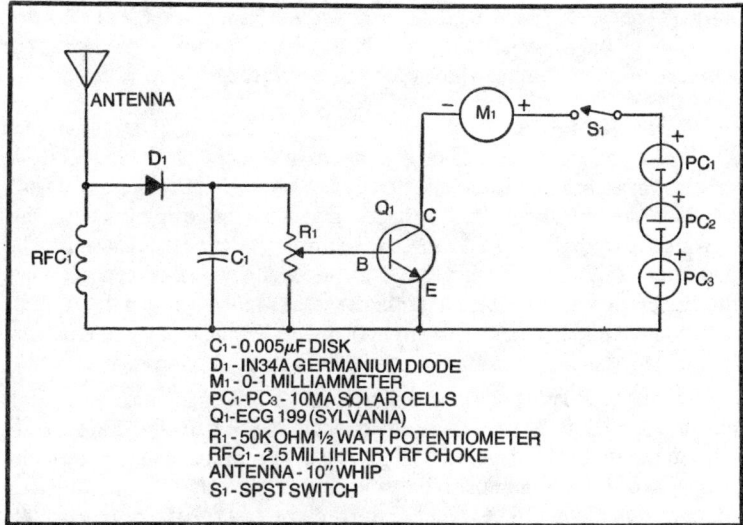

Fig. 4-74. Light-powered field-strength meter circuit.

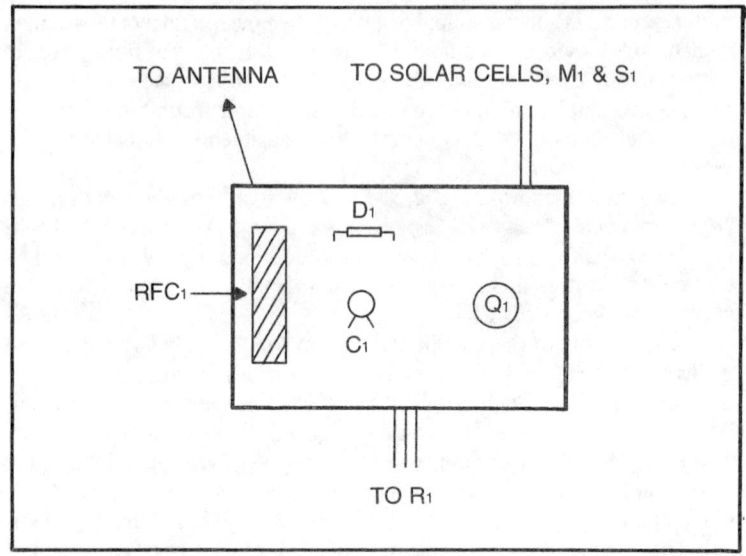

Fig. 4-75. Field-strength meter components on circuit board.

not especially critical, although component leads should be as short as possible. Referring to Fig. 4-76, the circuit is installed in a mini-box; one which will fit in the palm of your hand. The meter is mounted through the front of the metal enclosure, as is the variable control. The solar cells are attached to the top of the mini-box with another piece of perforated circuit board which has been epoxied to this location. The antenna is installed through the case. A small ceramic insulator is attached through the box for the base of the antenna to rest upon. A rubber grommet insulates the antenna from the aluminum box when it passes through to the outside.

Check-Out Procedure. With S1 in the off position, place the circuit and enclosure on a bench or other stationary object, directing the solar cells toward a bright source of light. Turn S1 to the on position and adjust R1 for a zero reading. This last step may not be necessary. Key the transmitter whose output is to be measured and adjust R1 again for a mid-scale reading on the meter. The various alignment procedures may then be carried out the transmitter under test while observing the results of the output power on the field-strength meter. When a maximum reading is obtained, the transmitter is putting out as much power as possible.

There are many polarized devices in this circuit, and the reversal of any one of them will cause this field-strength meter to malfunction. Make certain the meter is installed with its negative terminal connected to the collector of Q1 and tis positive terminal to the positive contact of PC1. Most malfunctions will be traced to the reversal of a polarized component, a defective component, or an inadequate source of light.

This field-strength meter project is designed to be used only when stationary. It will behave erratically if carried from site-to-site while observing a transmitter output reading. If it's necessary to constantly move this instrument, you may wish to install a 1.35 VDC rechargeable battery in the circuit in parallel with the solar cell string. You will then have a battery-operated field-strength meter which is charged by the sun.

LIGHT-CONTROLLED ROTARY SWITCH

Sometimes it becomes desirable to control many functions from one switching device. The circuit in Fig. 4-77 can control up to twelve switching functions using the stepping relay specified. The source of control is a transistor switch whose conduction is determined by PC1, a photoresistor. The actual electronic switching circuit is quite conventional and similar to other switching devices in this book. For this reason, this portion of the circuit will not be discussed as much as the stepping relay, K2, which is the most unusual part of the entire circuit.

K2 is a pulse-operated device very similar in operation to the ratchet relay used in an earlier project. Each time K1 closes, K2 advances to the next position. When K1 opens again, K2 holds the last position until K1 is closed again. When this occurs, the next switching position is contacted. This is a one-way device, so if you pass by the position you desire, it will be necessary to advance the relay eleven more positions, allowing the contact arm to rotate 360 degrees to the contact you have selected.

K2 is powered from a 115-volt ac source. K1 is powered by a 9-volt battery and switches power on and off to K2. R2 allows the switch to be

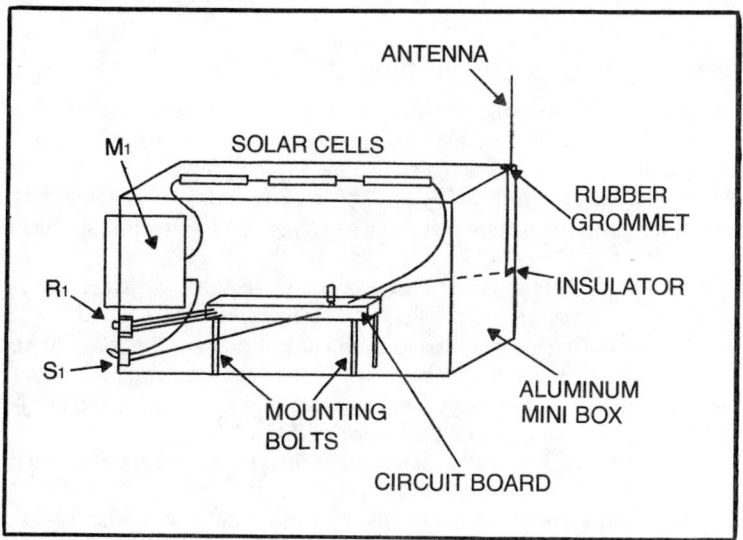

Fig. 4-76. An aluminum case serves as an enclosure for the circuit board and as a mount for the solar cells.

163

adjusted for various ambient light conditions. The electronic portion of the circuit and K1 are installed on a small piece of perforated circuit board in a similar manner to earlier projects. K2 is mounted separately in a plastic or aluminum enclosure side by side with the wired circuit board. Normally, K2's contacts will be brought out to a 12-contact terminal strip. A separate contact serves as the output terminal for the rotor portion of this stepping relay.

Check-Out Procedure. With S2 in the off position and S1 switched on, adjust R2 so that K1 triggers when the beam of a flashlight strikes its surface. Make certain K1 is not keyed by the ambient light in the room where this circuit is located. When the light is removed from PC1, K1 should disengage. Now, with the light removed from PC1, close S2. Nothing should occur at this point; but if K2 advances, R2 is not properly adjusted or K1's contacts have been wired to the K2 supply lines in the normally closed position. The normally open contacts of K1 are used for this circuit.

If K1 is triggering properly, direct the beam of the flashlight onto PC1. S2 should advance one step. Remove the light source and then shine the light on PC1 again. K2 should advance another step.

If the electronic circuit does not seem to be operating properly, check your connections to the transistor and to B1. Make certain no component leads are reversed. Also, make sure S1 is in the closed position. If the electronic circuit is triggering but K2 does not advance, the 115VAC power source may be disconnected or there may be a broken wire at K1's contacts.

VARIABLE-VOLTAGE SOLAR POWER SUPPLY

With the wide range of electronic circuits on today's market, it is often necessary to have a source of variable voltage to supply operating current at the correct potentials. Common DC supply voltages are 1.5, 3.0, 4.5, 6.0, 9.0, and 12.0 VDC. It is possible to build one complex solar cell supply which will deliver all of these voltages. The circuit is shown in Fig. 4-78.

A total of 28 solar cells are used for this project. No particular type is specified in the schematic drawing because the builder will have to decide which devices will be used based upon current drain of the loads to be powered. If the supply must deliver only 20 or 30 milliamperes, this entire circuit may be built for about $40; but if an output of up to 1 ampere is needed, this same supply could cost nearly $300.

A rotary switch is used to select the various output voltages. Looking at the schematic, when S1 is in the 1.5VDC position, only 4 of the 28 solar cells are used. The number of cells used progresses up to the 12-volt position, where all 28 are active in the circuit. The current rating of S1's contacts will be determined by the current output of the solar cells used. If

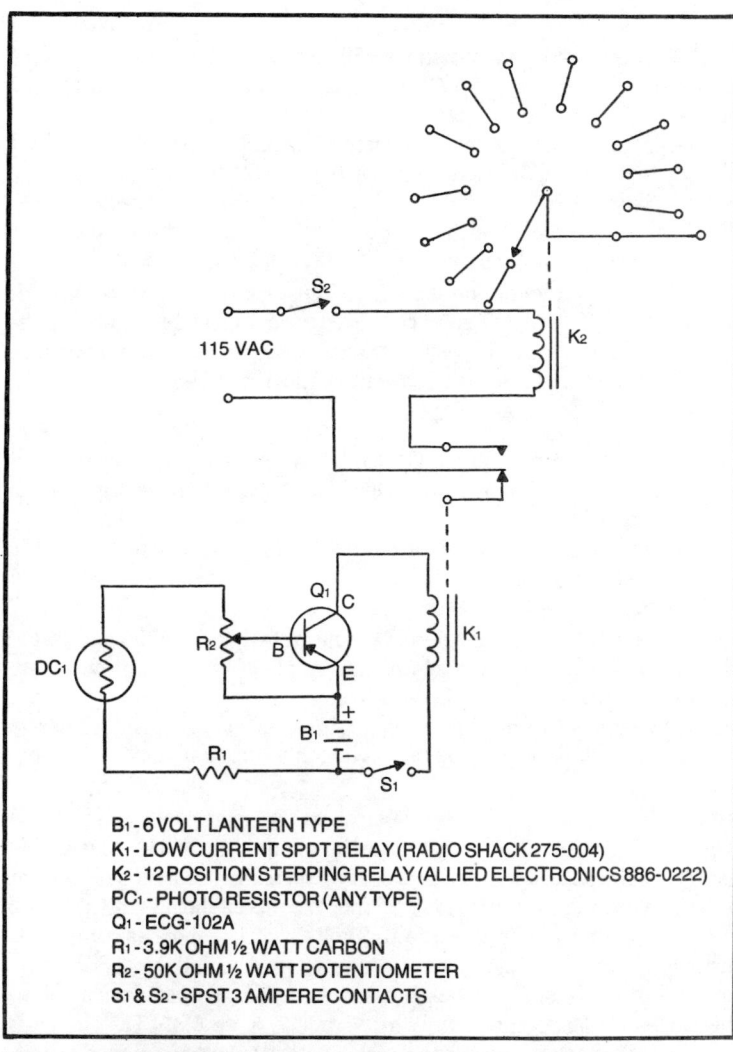

Fig. 4-77. Light-controlled rotary switch circuit.

1 ampere cells are selected, the switch should have contact ratings of at least 1 ampere. Lower-output cells will allow you to decrease the ratings of the switch.

D1 is a 50PIV, 1-ampere diode, but this rating should be increased to 1.5-or 2.0 amperes if 1-ampere solar cell devices are used. The 50PIV rating will still remain the same. If this circuit is never to be used as a battery charger, D1 may be omitted; but this is a safety feature which is very inexpensive and could save $300.00 worth of solar cells from destruction. Common sense suggests that this diode should be used.

It is impossible to provide specific construction information for this project, because these instructions will vary depending upon the size of the cells the builder selects. A low-current unit can easily be constructed on a large section of perforated circuit board; but a high-current model will require the use of a large, non-conductive panel. The diode may be wired directly to the circuit board or panel and the connective wiring brought to a remote control box where the switch is located. For low-current models, the circuit board containing the solar cells may be mounted atop a plastic or aluminum chassis with the switch and diode contained inside.

This is an excellent project for field uses of electronic devices. If moderate or high-current solar cells are chosen, low-powered transmitters and receivers of different operating voltages may be powered in remote sites where there is no conventional source of power.

Check-Out Procedure. Once the solar panel is built, connected all wires to the switch contact, as is indicated in the schematic drawing. An SPST toggle switch may also be located in the positive lead coming from S1 to allow you to switch current on and off without having to remove the panel from the rays of the sun. Be sure D1 is installed as shown. If you reverse this component, no current will flow.

Place a voltmeter capable of reading from 1 to 13 volts across the positive and negative terminals of the supply. For this portion of the check-out procedure, it may be best to connect the switch contacts to the solar cell segment with alligator-clip leads. After the exact voltage points are determined, these may be replaced with permanent soldered hookup wires.

With S1 in the 12-volt position, place the solar panel in a position where it will receive bright light from the sun or an artificial source. This panel must be positioned so that every cell receives a nearly equal amount of light. With the switch in this position, the meter should read about 12 volts. This value may be closer to 12.3 volts, which is about as close to the nominal 12-volt value as can be obtained using solar cells. If this value is lower than 12 volts, it may be necessary to add an additional solar cell to the series string. Some cells do not produce quite 0.45VDC, which is considered to be the average value of the voltage output of common units.

Turn S1 to the 1.5VDC position and measure the voltage on the meter. It should equal 1.55 volts if standard cells are used. All of these measurements assume that D1 is in the circuit. If this component is omitted, output voltage will be slightly higher. Continue rotating S1, checking all of the voltage levels to make certain they conform with the stated ratings of the schematic. If alligator-clip leads are used as suggested, you may alter the connections to the solar cell string to arrive at the desired voltage output.

If you get absolutely no reading on the meter, you probably have a broken wire, defective switch, or a reversal of D1. You may also have a

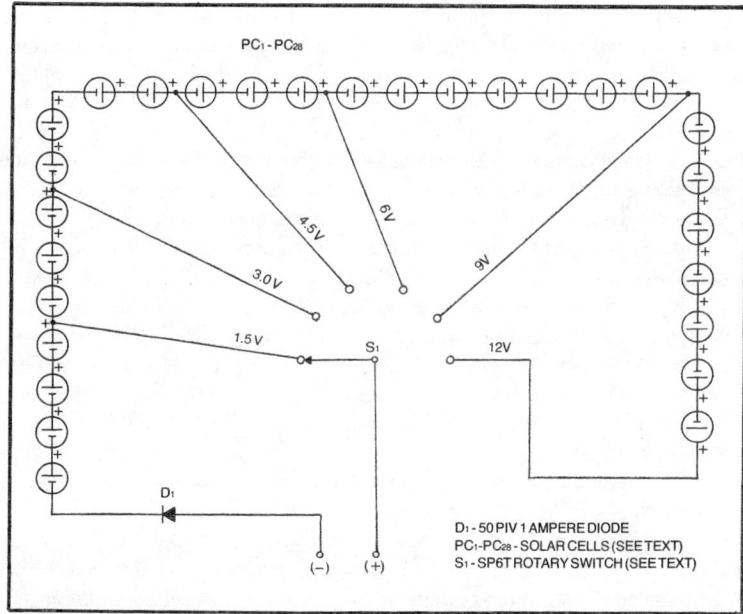

Fig. 4-78. Variable-voltage solar power supply circuit.

reversal of one of the solar cells, probably one of the first ones in the string, using the recommended test procedure. If the 1.5, 3.0, and 4.5 VDC contacts produce the desired voltage outputs, but the 6.0 VDC contact produces nothing, this is an indication of a defective or reversed solar cell between the 4.5 VDC connection to the string and that of the 6-volt connection. If one cell is reversed or defective, no output will be obtained when the switch is rotated past that point.

With this solar power supply, you have a great deal of versatility in providing operating current to many different low-powered electronic circuits. If miniature solar cells are used, this circuit becomes very portable and may be taken to remote locations and on field-day outings.

SOLAR-POWERED TELEVISION RECEIVER

The last project in this book is somewhat expensive and quite unusual. It involves powering a television receiver directly from sunlight. There are some receivers around which contain internal batteries which can be charged by sunlight, but these are only solar *assisted*. They are powered from their battery packs which are charged by solar energy; however, they do not operate directly from converted solar power.

The secret of this project is not so much in the solar power supply end of it as it is in choosing the television receiver. The author wished to keep costs to a minimum regarding the solar supply, so the Sinclair Micro Television receiver was selected. Manufactured by Sinclair Radionics in England, it is advertised as the smallest television in the world and has been featured on the Guiness Book of World Records television program. Shown with the solar power supply in Fig. 4-79, it can be seen that this receiver easily fits in the palm of the hand and is designed for "close-in" viewing.

This particular model originally sold for nearly $400, which is quite expensive for a black and white television receiver; however, J S & A Products That Think, in Northbrook, Illinois, markets this device in the United States for less than $250.00. One of the reasons for choosing this particular model was its low power requirement. It may be powered from a 12-volt automotive system, internal 12-volt batteries, AND from a 6VDC source at a little less than 400 milliamperes. This 6VDC power requirement was the most interesting factor about the set when considering solar power because almost every other battery-operated television requires a 12 VDC source. Half the number of solar cells could be used in supplying 6 VDC.

It should be stated at this point that $250 may still be a large sum of money to pay for a black and white television receiver, but this one is so interesting and can be used for many purposes and in many remote places. While there are larger but less expensive battery operated portables on the market, the additional cost in solar cells would more than outweigh any savings in purchase price. Admittedly, the author did not have to pay for the television set he used. Joe Sugarman, the owner of J S & A, became so interested in this project that he sent the receiver on a gratis basis. The reader may not be quite this fortunate, but the idea of the microtelevision is not as new today as it was when this one was obtained, so you may be able to pick one up on sale for around $150.

Fig. 4-79. The Sinclair MicroTelevision is easily held in one hand. Its solar power supply is visible in the background.

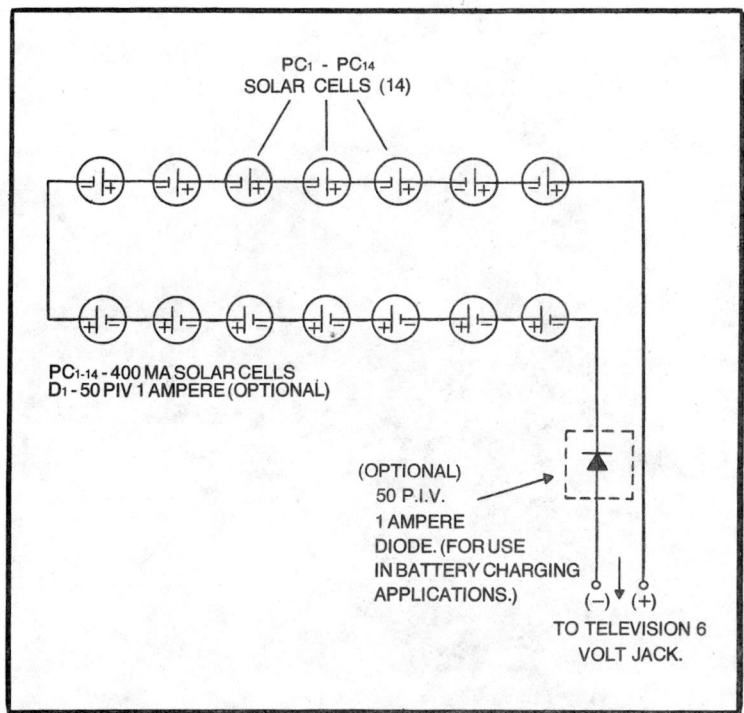

Fig. 4-80. Schematic diagram of solar power supply circuit.

Figure 4-80 shows the schematic of the 6-volt power supply. An optional protective diode is shown; however, this was not used in the author's original supply. Fourteen solar cells provide an output of a little over 6 VDC, which is adequate for fully powering the receiver. Originally, 400-milliampere Radio Shack cells were used. They were purchased for about $5.50 each, but these are no longer stocked and have been replaced by 1-ampere units that sell for about $10 each. These units will enable the set to operate under less than ideal lighting conditions but the expense may be a little steep for some experimenters. It should be possible to pick up solar cells with minimum current output of 400 milliamperes on the industrial surplus market. Allied Electronics offers 550-milliampere units in quantities of 10 or more for $8.90 each. Since the television draws less than 400 milliampers at 6 VDC, the current supplied by these units is more than adequate.

Almost every type of cell purchased for this application will come without leads. These must be soldered on after purchase. Figure 4-81 shows how the negative lead is connected to the front of the solar cell. This is a delicate process, as the cells are quite fragile until mounted. Several of them cracked during the process but were mended by placing scotch tape across the back to re-bond the two sections. It is also necessary to

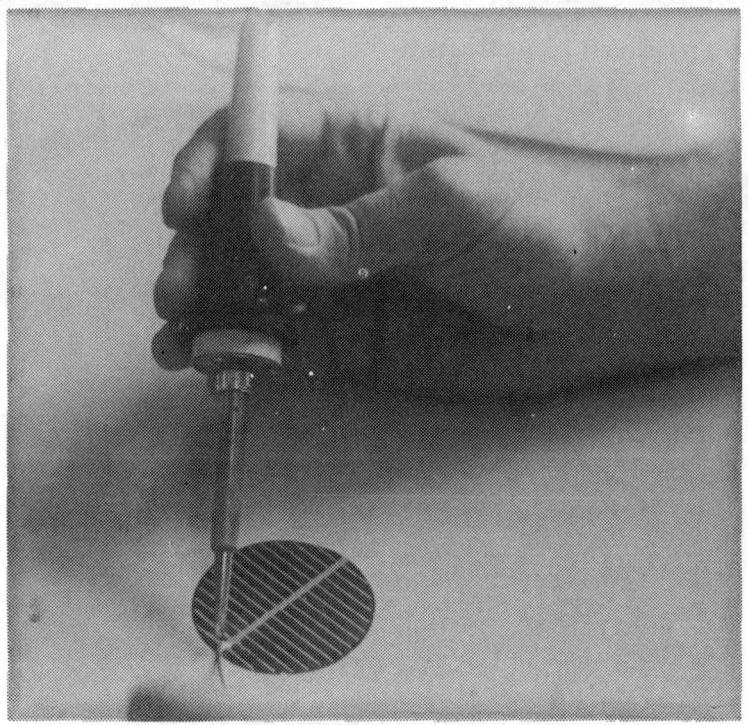

Fig. 4-81. Component lead is soldered to front strip of solar cell.

electrically connect the two sections by bridging the break with a small conductor soldered to both pieces.

Figure 4-82 shows how the solar cells were mounted. Circular foam pads were glued to a piece of thick cardboard measuring 13 x 9 inches. This in turn was glued to a 14 x 10-inch section of ¼-inch plywood. The entire assembly was mounted in a small wooden box. This was an experiment, so the solar cells were not permanently bonded to a large section of plexiglass. This sloppy but simple mounting method allowed the changing of wiring connections and the rearranging of the cells. Figure 4-83 shows the completed project. It is obvious that this is a temporary arrangement, but the electrical operation is not hindered in any way. After attaching the leads to the cells, they were glued to the foam rubber pads and temporarily arranged in the manner shown in Fig. 4-82. From testing the output voltage, it was determined that the particular cells chosen delivered exactly 6.0 volts when thirteen of them were connected in series. Fourteen cells are recommended if the optional diode is used.

When the glue is set, the excess writing from cell-to-cell was forced through holes in the cardboard and secured by epoxy cement. Plastic insulators were used to hold the wiring connections flat on the cardboard to prevent them from causing shadows across the faces of the cells. The 6-volt power cord was connected to the output of this solar supply, and all was ready for the final test.

At this point, the project seemed to have gone too easily. A total time of about three hours was required for completion of this project. It is at times like these that a builder may feel that anything this easy just can't work the first time. In this case, I was right. The television was tuned to a nearby station using the internal 12-volt battery supply; with the solar bank placed in the sun on a perfect cloudless afternoon, the power plug was inserted in its 6-volt socket; the receiver immediately went dead. Removing the plug caused the television to become operable again from its internal batteries. The output from the solar supply was checked again and read exactly 6.0 volts. The plug was inserted again and the television still refused to work. It was only then that the author discovered that the plug was not completely entering its receptacle due to an extraneous piece of plastic around the mouth. This was quickly removed, the plug inserted fully, and the receiver operated exactly as if it were still being powered from the internal batteries. To make certain that the solar supply was actually powering the set, a piece of cardboard was used to cast a shadow on the solar cells. The receiver ceased to function. When the shadow was removed, it was immediately operational again. Figure 4-84 shows the television which is dwarfed by the solar cell supply.

Solar-derived electricity is exactly like battery power as far as the television receiver is concerned. Absolutely no difference in picture or audio quality was noted. As a matter of fact, the operation of the set was quite typical and no exciting or exotic discoveries were made. Later tests have shown that the receiver can be powered from the solar cells even when the sun is behind thin clouds. If a glare can be obtained on the cell faces, adequate current is present to drive the set. When light falls below

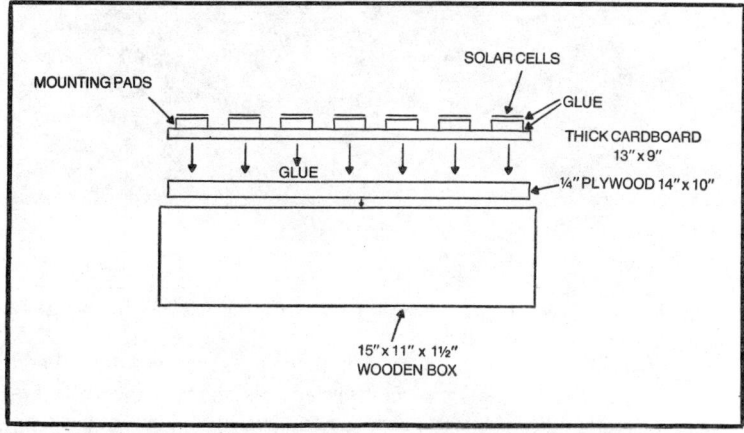

Fig. 4-82. Mounting of solar cells on rigid platform for encasement in wooden box.

Fig. 4-83. Completed mock-up of solar cell supply.

acceptable levels, the picture will drop out first, followed by the audio as light levels continue to diminish.

The reader, by no means, has to adhere to the exact construction used by the author for this experimental project. As long as the cells are properly wired and connected, almost any mounting configuration can be used which allows them to be exposed to bright light sources without shading. As an added note, any device which requires 6 volts of operating current at levels equal or to less than the current ratings of the cells can be as efficiently powered.

Any battery-operated television can be powered by a solar cell supply. Most, however, will require 12-volt operation which would effectively double the number of solar cells used, although current requirements might be reduced, depending upon the set.

While this was an experiment to see if a television receiver could be efficiently powered directly from the sun, this project resulted in a very practical use. The solar cell supply was taken along on many camping outings where no conventional sources of power were available. Solar power was used to drive the set in the daytime. The internal batteries were used at night. It would even be possible to provide enough operating current from the flames of a roaring campfire. Unfortunately, this requires that the solar supply be placed in too close a proximity to the fire. The solar cells would quickly be destroyed by the heat.

With this project, you can watch your favorite programs anywhere the sun is shining; but at night, you either switch to internal batteries, an alternate power supply, or do without until the sun rises again. For the hiker and camper who takes long excursions into the countryside, this may be a very worthwhile project which will provide many hours of television viewing far from the comforts of home.

SUMMARY

The projects contained in this chapter are designed to be informational. Most are also useful in certain applications. Some are just plain fun,

Fig. 4-84. Microtelevision is dwarfed by solar power supply.

while still providing the educational benefits of experimenting with and building a specialized circuit in the home workshop.

The reader is encouraged to go further and try combining certain projects, adding to them, and making modifications where necessary. In this way, you may arrive at a new circuit based upon old ideas and schematics which will do a job more efficiently. Once you become accustomed to them, light-sensitive solid-state devices are quite easy to work with and offer wide variations in application and theory.

When building future electronic projects, you may see ways that these can be adapted to solar power or light actuation and control. Through these experiments, you will become more knowledgeable in the many uses of solar cells, photoresistors, phototransistors, and all of the other devices and circuits which depend upon light energy to operate.

Appendix A
Schematic Symbols

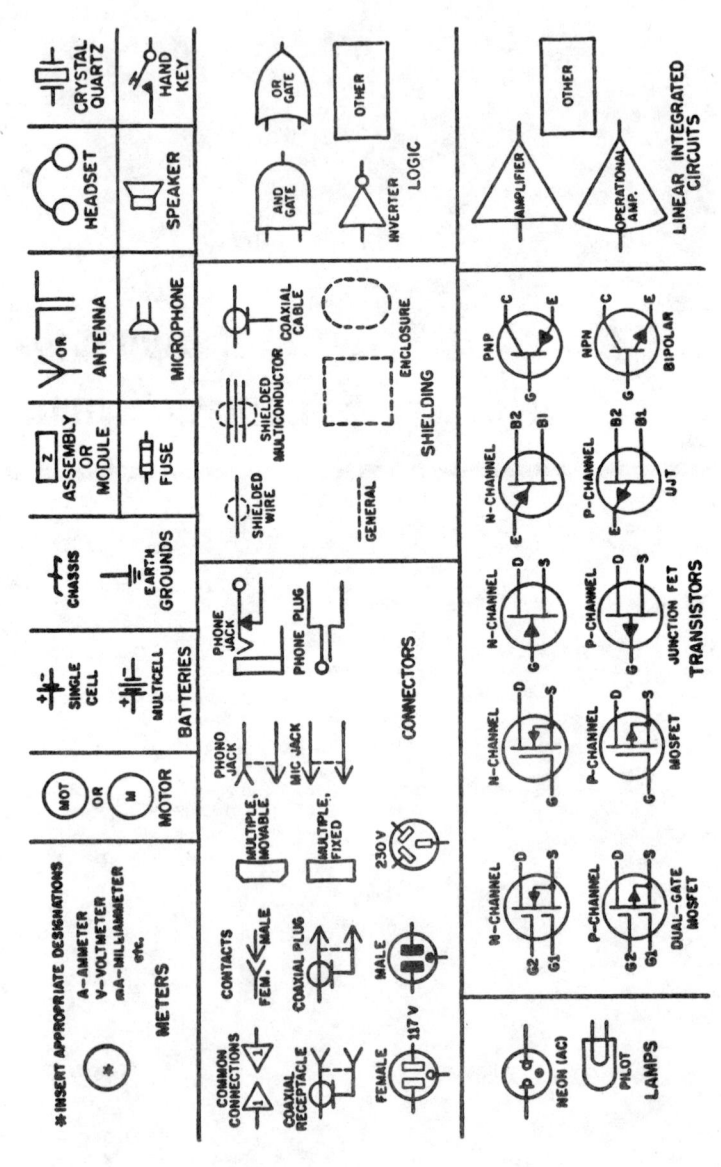

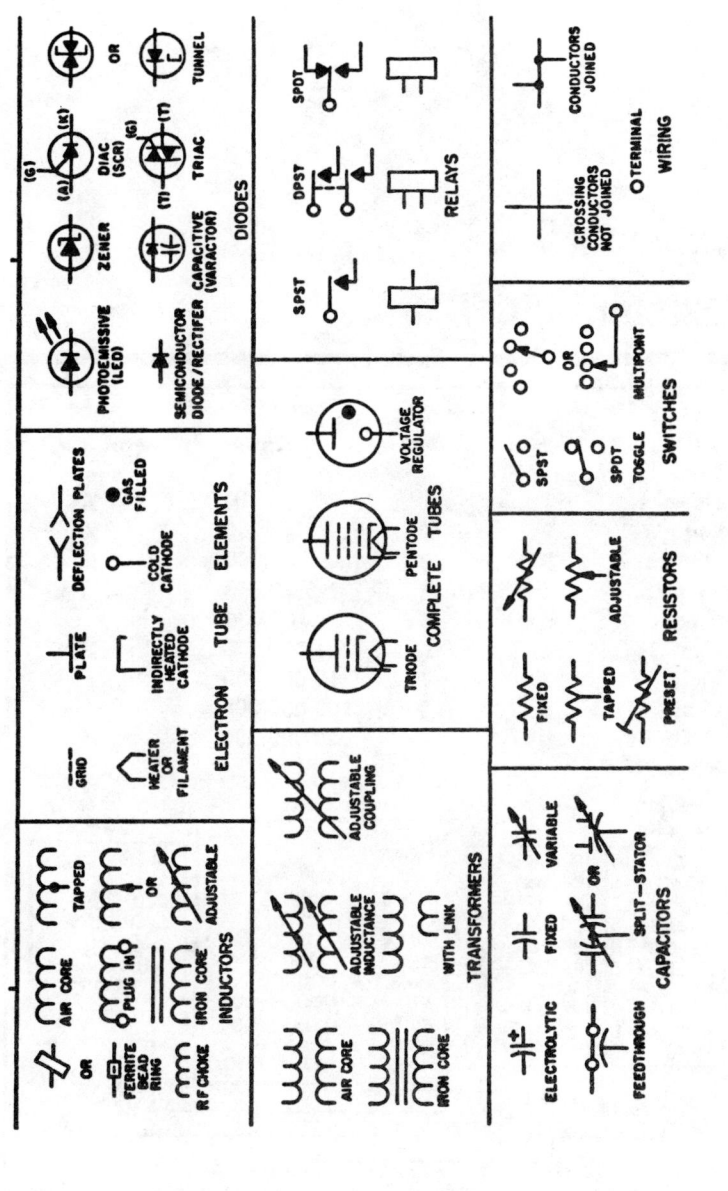

Appendix B
Resistor Color Chart

Color	First Digit	Second Digit	Multiplier	Tolerance (±)
Black	—	0	1	—
Brown	1	1	10	1%
Red	2	2	100	2%
Orange	3	3	1,000	3%
Yellow	4	4	10,000	4%
Green	5	5	100,000	
Blue	6	6	1,000,000	
Violet	7	7	10,000,000	
Gray	8	8	100,000,000	
White	9	9	1,000,000,000	
Silver				10%
Gold				5%
No color				20%

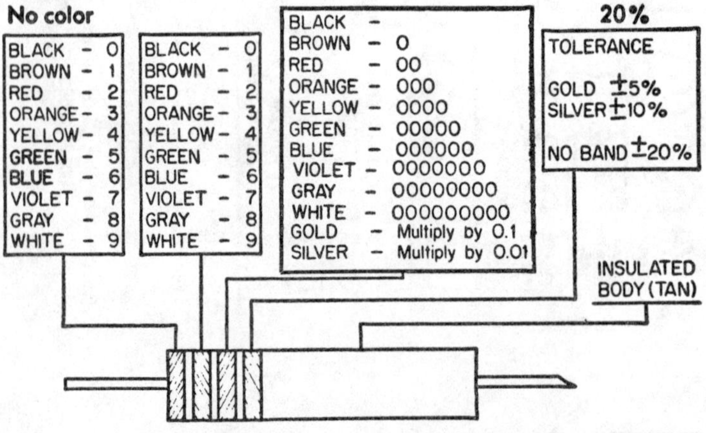

```
BLACK  - 0      BLACK  - 0
BROWN  - 1      BROWN  - 1
RED    - 2      RED    - 2
ORANGE - 3      ORANGE - 3
YELLOW - 4      YELLOW - 4
GREEN  - 5      GREEN  - 5
BLUE   - 6      BLUE   - 6
VIOLET - 7      VIOLET - 7
GRAY   - 8      GRAY   - 8
WHITE  - 9      WHITE  - 9
```

```
BLACK  -
BROWN  - 0
RED    - 00
ORANGE - 000
YELLOW - 0000
GREEN  - 00000
BLUE   - 000000
VIOLET - 0000000
GRAY   - 00000000
WHITE  - 000000000
GOLD   - Multiply by 0.1
SILVER - Multiply by 0.01
```

TOLERANCE

GOLD ±5%
SILVER ±10%

NO BAND ±20%

INSULATED BODY (TAN)

Index

A

AC devices, high-current control	138
Alarm clock, electronic	96
electronic check-out procedure	98
AM radio booster	101
check-out procedure	102
AM radio, solar-powered	107
solar-powered check-out procedure	108
Amateur radio transmitter	112
check-out procedure	114
Audible light meter	121
check-out procedure	122
Automobile finder, light-controlled	142
light-controlled check-out procedure	146

B

Battery charger, 12-volt	109
check-out procedure	111
Booster, radio AM	101
check-out procedure	102
Building habits	59
Building tools	50

C

Cadmium selenium cell	26
Cadmium sulfide cell	16
CdS cell	16
CdSe cell	26
Cell, cadmium selenium	26
cadmium sulfide	16
CdS	16
CdSe	26
photoconductive	15
photoelectric	15
photovoltaic	15, 17
Cells, photoconductive	26
solar	15
solar combining	21
solar shock-mounting	61
solar special techniques	61
Charger, battery 12-volt	109
check-out procedure	111
Check-out procedure,	
amateur radio transmitter	114
AM radio booster	102
audible light meter	122
code practice oscillator	117
combination light meter	86
electronic alarm clock	98
high-current SCR switch	137
light-controlled automobile finder	146
light-controlled electronic organ	99
light-controlled holding switch	160
light-controlled rotary switch	164
light modulator	147
light-powered/	

light-controlled pulse oscillator	126
light-powered field-strength meter	162
light-regulated power supply	152
phototransistor commercial killer	119
photoresistor light meter	84
photovoltaic light meter	81
remote light-controlled potentiometer	153
six-volt solar power supply	92
solar-powered AM radio	108
solar-powered FM radio	105
solar-powered 2-meter converter	134
solar-powered RF oscillator	142
solar-powered watch	130
three-volt solar power supply	88
variable-voltage solar power supply	166
versatile light-controlled switch	94
Circuit, extension relay	95
integrated	64
integrated building techniques	64
parallel	23
series	21
Code practice oscillator	115
oscillator check-out procedure	117
Coherent light	13
Combination light meter	85
check-out procedure	86
Commercial killer, phototransistor	118
phototransistor check-out procedure	119
Components, electronic keeping track of	72
obtaining	68
solid-state mounting	63
Concentrating the light	20
Converter, solar-powered 2-meter	133
check-out procedure	134
Cross-referencing	70

D

DIP	64
Dual in-line package	64

E

Electromagnetic spectrum	9
Electronic alarm clock	96
check-out procedure	98
Electronic components, keeping track of	72
Electronic organ, light-controlled	98
check-out procedure	99
Electronic project building	41
Electronic solar projects	78
Energy	9
Experimenter's junk box	70
Extension relay circuit	95

F

Field-strength meter, light-powered	160
check-out procedure	162
FM radio, solar-powered	103
check-out procedure	105
Frequency	7

H

High-current control of AC devices	138
High-current SCR switch	135
check-out procedure	137
High-current supply, six-volt	90
Holding switch, light-controlled	157
check-out procedure	160
Horizontal mounting	47

I

IC	64
Instruments, test	53
Integrated circuit	64
building techniques	64

J

Junk box, experimenter's	70

L

LASCR	39
Lasers	12
Light	8
Light-activated silicon controlled rectifier	39
Light, coherent	13
concentrating	20
Light-controlled automobile finder	142
check-out procedure	146
Light-controlled electronic organ	98
check-out procedure	99
Light-controlled holding switch	157
check-out procedure	160
Light-controlled	

potentiometer, remote	153
remote check-out procedure	156
Light-controlled rotary switch	163
check-out procedure	164
Light-controlled switch, versatile	93
check-out procedure	94
Light, makeup of	7
Light meter, audible	121
check-out procedure	122
Light meter, combination	85
check-out procedure	86
Light-meter, photoresistor	82
check-out procedure	84
Light meter, photovoltaic	80
check-out procedure	81
Light modulator	146
check-out procedure	147
Light-powered	
field-strength meter	160
check-out procedure	162
Light-powered/	
light-controlled	
pulse oscillator	123
check-out procedure	126
Light-regulated power supply	150
check-out procedure	152
Light-sensitive devices, other	39

M

Makeup of light	7
Meter, audible light	121
check-out procedure	122
Meter, combination light	85
check-out procedure	86
Meter, light-powered	
field-strength	160
check-out procedure	162
Meter, photoresistor light	82
check-out procedure	84
Meter, photovoltaic light	80
check-out procedure	81
Modulator, light	146
check-out procedure	147
Motorized pinwheel	149
Mounting, horizontal	47
Mounting of solid-	
state components	63
Mounting, vertical	48

O

Opaque	9
Oscillator, code practice	115
check-out procedure	117
Oscillator, light-powered/	
light-controlled pulse	123
check-out procedure	126

Oscillator, solar-powered RF	139
check-out procedure	142

P

Parallel circuit	23
Perf board	41
Perforated circuit board	41
Pinwheel, motorized	149
Photoconductive cell	15
Photoconductive cells	26
Photoelectric cell	15
Photoelectric effect	10, 17
Photoresistor light meter	82
check-out procedure	84
Phototransistors	36
Phototransistor commercial killer	118
check-out procedure	119
Photovoltaic cell	15, 17
Photovoltaic light meter	80
check-out procedure	81
Potentiometer, remote	
light-controlled	153
check-out procedure	156
Power supply, light-regulated	150
check-out procedure	152
Power supply, 150-volt	131
check-out procedure	132
Power supply, six-volt solar	90
check-out procedure	92
Power supply, three-volt solar	86
check-out procedure	88
Power supply, variable-	
voltage solar	164
check-out procedure	166
Project building, electronic	41
Projects, electronic solar	78
Pulse oscillator,	
light-powered/	
light-controlled	123
check-out procedures	126

R

Radio, AM solar-powered	107
check-out procedure	108
Radio booster, AM	101
check-out procedure	102
Radio, FM solar-powered	103
check-out procedure	105
Receiver, television	
solar-powered	167
Remote light-controlled	
potentiometer	153
check-out procedure	156
RF oscillator, solar-powered	139
check-out procedure	142
Rotary switch, light-controlled	163
check-out procedure	164

S

SCR switch, high-current	135
check-out procedure	137
Series circuit	21
Shock-mounting of solar cells	61
Six-volt high-current supply	90
Six-volt solar power supply	90
check-out procedure	92
Solar cells	15
combining	21
shock-mounting	61
special techniques	61
Solar-powered AM radio	107
check-out procedure	108
Solar-powered FM radio	103
check-out procedure	105
Solar-powered RF oscillator	139
check-out procedure	142
Solar-powered television receiver	167
Solar-powered 2-meter converter	133
check-out procedure	134
Solar-powered watch	126
check-out procedure	130
Solar power supply, six-volt	90
check-out procedure	92
Solar power supply, three-volt	86
check-out procedure	88
Solar power supply, variable-voltage	164
check-out procedure	166
Solar projects, electronic	78
Soldering procedures	54
Soldering techniques	56
Spectrum, electromagnetic	9
Summary, electronic project building	76
Summary, electronic solar projects	173
Summary, light	14
Summary, solar cells	40
Switch, holding light-controlled	157
check-out procedure	160
Switch, rotary light-controlled	163
check-out procedure	164

T

Television receiver, solar-powered	167
Test instruments	53
Three-volt solar power supply	86
check-out procedure	88
Tools	50
Transducers	15
Transistors	30
Translucent	9
Transmitter, amateur radio	112
check-out procedure	114
Transparent	9
Troubleshooting	79

V

Variable-voltage solar power supply	164
check-out procedure	166
Versatile light-controlled switch	93
check-out procedure	94
Vertical mounting	48
Visible radiant energy	8
Voltage curve	7

W

Watch, solar-powered	126
check-out procedure	130
Wavelength	7